Assessing Revolutionary and Insurgent Strategies

CASE STUDIES IN INSURGENCY AND REVOLUTIONARY WARFARE— THE PATRIOT INSURGENCY (1763–1789)

Robert R. Leonhard, Lead Author
Johns Hopkins University Applied Physics Laboratory (JHU/APL)

JHU/APL Contributing Authors

Summer D. Agan

Stephen P. Phillips

United States Army Special Operations Command

Case Studies in Insurgency and Revolutionary Warfare—The Patriot Insurgency (1763-1789) is a work of the United States Government in accordance with Title 17, United States Code, sections 101 and 105.

Cite me as:

Leonhard, Robert, et al. *Case Studies in Insurgency and Revolutionary Warfare: The Patriot Insurgency (1763-1789)*. Fort Bragg: US Army Special Operations Command, 2019.

Published by Conflict Research Group.

First published by USASOC in 2019

This edition © Conflict Research Group, 2024, all rights reserved.
Cover art © Conflict Research Group, 2024, all rights reserved.

No part of this publication may be reproduced, stored in a retrieval system or transmitted, in any form or by any means, electronic, mechanical, photocopying, recording or otherwise without prior permission from the publisher.

Reproduction in whole or in part is permitted for any purpose of the United States government. Nonmateriel research on special warfare is performed in support of the requirements stated by the United States Army Special Operations Command, Department of the Army. This research is accomplished at the Johns Hopkins University Applied Physics Laboratory by the National Security Analysis Department, a nongovernmental agency operating under the supervision of the USASOC Sensitive Activities Division, Department of the Army.

The analysis and the opinions expressed within this document are solely those of the authors and do not necessarily reflect the positions of the US Army or the Johns Hopkins University Applied Physics Laboratory.

Comments correcting errors of fact and opinion, filling or indicating gaps of information, and suggesting other changes that may be appropriate should be addressed to:

United States Army Special Operations Command

G-3X, Sensitive Activities Division

2929 Desert Storm Drive

Fort Bragg, NC 28310

All ARIS products are available from USASOC at www.soc.mil under the ARIS link.

ASSESSING REVOLUTIONARY AND INSURGENT STRATEGIES

The Assessing Revolutionary and Insurgent Strategies (ARIS) series consists of a set of case studies and research conducted for the US Army Special Operations Command by the National Security Analysis Department of the Johns Hopkins University Applied Physics Laboratory.

The purpose of the ARIS series is to produce a collection of academically rigorous yet operationally relevant research materials to develop and illustrate a common understanding of insurgency and revolution. This research, intended to form a bedrock body of knowledge for members of the Special Forces, will allow users to distill vast amounts of material from a wide array of campaigns and extract relevant lessons, thereby enabling the development of future doctrine, professional education, and training.

From its inception, ARIS has been focused on exploring historical and current revolutions and insurgencies for the purpose of identifying emerging trends in operational designs and patterns. ARIS encompasses research and studies on the general characteristics of revolutionary movements and insurgencies and examines unique adaptations by specific organizations or groups to overcome various environmental and contextual challenges.

The ARIS series follows in the tradition of research conducted by the Special Operations Research Office (SORO) of American University in the 1950s and 1960s, by adding new research to that body of work and in several instances releasing updated editions of original SORO studies.

RECENT VOLUMES IN THE ARIS SERIES

Casebook on Insurgency and Revolutionary Warfare: Volume I: 1927–1962 (2013)
Casebook on Insurgency and Revolutionary Warfare: Volume II: 1962–2009 (2012)
Human Factors Considerations of Undergrounds in Insurgencies (2013)
Undergrounds in Insurgent, Revolutionary, and Resistance Warfare (2013)
Understanding States of Resistance (2019)
Legal Implications of the Status of Persons in Resistance (2015)
Threshold of Violence (2019)
"Little Green Men": A Primer on Modern Russian Unconventional Warfare, Ukraine 2013–2014 (2015)
Science of Resistance (forthcoming)

TABLE OF CONTENTS

CHAPTER 1. INTRODUCTION AND SUMMARY 1
    Background 3
    Purpose of the Case Study 4
    Organization of the Study 6
    Methodology of the Study 7
        Physical Environment 8
        Historical Context 9
        Socioeconomic Conditions 11
        Government and Politics 13
    Synopsis of Case Study in Insurgency and Revolutionary Warfare:
        Patriot Insurgency (1763–1789) 15
    Timeline 15

PART I. CONTEXT AND CATALYSTS OF THE INSURGENCY 23

CHAPTER 2. PHYSICAL ENVIRONMENT 25
    The Geography of a Revolution 27
        Time and Distance 28
        Infrastructure 28
        Key Regions 29
            Ports 29
            Hudson Valley-Lake Champlain Coridor and Canada 29
            Mohawk Valley and the Great Lakes 30
            New England 31
            Middle Colonies 31
            Southern Colonies 32
            The West Indies 33

CHAPTER 3. HISTORICAL CONTEXT 35
    Introduction 37
    The American Colonies 37
        South Carolina 37
        North Carolina 38
        Georgia 38
        Virginia 38
        Maryland 39

  Pennsylvania 39
  Delaware 40
  New Jersey...40
  New York...41
  Massachusetts Bay..41
  Rhode Island 42
  New Hampshire 42
  Connecticut...42
 Native Americans..43
 The French and Indian War... 69
  From Victory to Crisis 69

CHAPTER 4. SOCIOECONOMIC CONDITIONS 81
 Population 83
  Social Classes.. 83
  Religion 85

CHAPTER 5. GOVERNMENT AND POLITICS 89
 British North America 91
 British West Indies 91
  Jamaica and the Sugar Colonies 91
  British Policy and the American Colonies........................... 92
   The Sugar Act of 1764.. 92
   The Currency Act of 1764.. 93

PART II. Structure and Dynamics of the Insurgency 97

CHAPTER 6. PATRIOT INSURGENCY 99
 The Course of the Patriot insurgency 101
  Pontiac's War 103
  The Stamp Act Crisis 104
  The Townshend Acts.. 108
  The Boston Massacre 110
  The Tea Act, Boston Tea Party, and Intolerable Acts 111
 Leadership, Organizational Structure, and Command
  and Control 117
 Underground Component and Auxiliary Component............. 117
 Armed Component.. 118
  The Nature of the Resistance.. 119
  Geography of the Theater of War...................................... 120
  Experience and Education of Patriot Officers.......................... 121

| Public Component 121
        Newspapers 121
    Ideology 123
        Religious Influences 123
        Good King, Bad King 125
        The Meaning of Liberty 126
        Conspiracy Theory 128
        This Land is My Land 129
        Choose a Side 129
        The Question of Independence 130
    Motivation and Behavior 131
        The Character of the Revolution 132
        Elites and Commoners 133
    Operations 134
        Paramilitary 134
            Land Operations 134
            Abortive Invasion of Canada, 1775 135
            The Campaigns in New England and the Middle States,
                1776-1778 137
            The Southern Campaigns, 1778-1783 141
            Analysis of Land Operations 144
        Naval Operations 144
            The Patriot Insurgency at Sea 144
            The Gaspee Affair 148
            The First Lake Champlain Maritime Campaign, 1775 149
            Washington's Navy, State Privateers 150
            Formation of a Naval Committee/Marine Committee 151
            External Support 153
            The Second Lake Champlain Campaign, 1776 153
            The Battle of Valcour Island 155
            Sustaining Foreign Assistance and Privateering 155
            Father of the American Navy 157
            French Naval Intervention 160
            The Battle of Ushant 160
            D'Estaing's Deployment 161
            Yorktown 163

CHAPTER 7. GOVERNMENT COUNTERMEASURES 169

Initial Response to Insurgency 171
From Political to Military Response 172

CHAPTER 8. CONCLUSION 175
    Transition from Insurgency to Governance 177
    The Unseen Seed of Rebellion 179
BIBLIOGRAPHY 181
INDEX 189

LIST OF ILLUSTRATIONS

Figure 2-1: The thirteen colonies. 27

Figure 2-2: The Hudson Valley-Lake Champlain corridor. 30

Figure 2-3: The southern colonies. 32

Figure 4-1: Population density, 1775. 84

Figure 6-1: George Grenville, prime minister, 1763-1765. 105

Figure 6-2: The *Pennsylvania Journal* 122

Figure 6-3: Patriot invasion of Canada, 1775. 136

Figure 6-4: The Battle of Long Island, retreat, and
            counterattack, 1776. 138

Credits

Figure 2-1. The Thirteen Colonies. Image from Nystrom Education; permission requested and received April 19, 2017.

Figure 2-2. The Hudson Valley-Lake Champlain Corridor. US Military Academy, *The Invasion of Canada, September 1775 - October 1776* [map] Department of History Maps, https://westpoint.edu/sites/default/files/inline-images/academics/academic_departments/history/Am%20Rev/07InvasionCanada.pdf.

Figure 2-3. The Southern Colonies. US Military Academy, *The War in the South, 1781* [map] Department of History Maps, https://westpoint.edu/sites/default/files/inline-images/academics/academic_departments/history/Am%20Rev/38South1781.pdf.

Figure 4-1. Population Density, 1775. US Military Academy, *Population Density, 1775* [map] Department of History Maps, https://westpoint.edu/sites/default/files/inline-images/academics/academic_departments/history/Am%20Rev/02PopulationDensity.pdf.

CHAPTER 1.
INTRODUCTION AND SUMMARY

The cause of America is in a great measure the cause of all mankind.

—Thomas Paine, *Common Sense*

BACKGROUND

The purpose of the Assessing Revolutionary and Insurgent Strategies (ARIS) series is to produce academically rigorous yet operationally relevant research to expand on and update the body of knowledge on insurgency and revolution for members of the US Special Forces. We began this work with a rigorous assessment of all known insurgent or revolutionary activities from 1962 through the present day. To conduct this assessment, we agreed on a basic definition of revolution or insurgency.[1,2] For the purpose of this research, a *revolution* is defined as:

> An attempt to modify the existing political system at least partially through unconstitutional or illegal use of force or protest.[3]

Next, we developed a taxonomy to establish a standard structure for analysis and to facilitate discussion of similarities and differences. We classified the events and activities according to the most evident cause of the revolt. The causes or bases of revolution were categorized as follows:

- Those motivated by a desire to greatly **modify the type of government**
- Those motivated by **identity or ethnic issues**
- Those motivated by a desire to **drive out a foreign power**
- Those motivated by **religious fundamentalism**
- Those motivated by **issues of modernization or reform**

After applying this taxonomy, we selected twenty-three cases, across the five previously stated categories, to be researched for inclusion in the *Casebook on Insurgency and Revolutionary Warfare Volume II: 1962–2009*.[4] For each of the twenty-three revolutions or insurgencies, the casebook includes a summary case study that focuses on the organization and activities of the insurgent group.

Subsequently, we selected several of the cases for a more detailed treatment that would apply a broader and more holistic analytical perspective, considering factors such as the social, economic, historical, and political context. Within the ARIS research series, these studies are referred to as "ARIS Tier 1 Insurgency Case Studies." This case study on the Patriot Insurgency is one of these works.

PURPOSE OF THE CASE STUDY

This case study examines the Patriot insurgency that developed among the English colonies in North America in the mid-eighteenth century and that eventually declared, fought for, and achieved independence from the mother country. There is a wealth of historical studies of the American Revolution, but this case study offers a unique perspective. Instead of simply repeating the well-documented history of the Revolution, we scrutinize the Patriots through the lens of modern insurgency doctrine and concepts.

As with every insurgency in human history, the Patriot insurgency had features that were similar to other resistance movements and features that were unique. Among the former were the political networking and negotiation that underlay the insurgency before it could coalesce and the difficulty in achieving and sustaining an armed component to carry out the wishes of the insurgent leaders. Among the latter, we can examine the part played by emerging and often conflicting ideological movements that proceeded from the Enlightenment, as well as the problem of warring against both Indians and British regulars backed by the most powerful navy in the world.

The Patriot insurgency wrestled with four major problems. First, there was the need to defeat or at least outlast the armies and navy of Great Britain. Second, the Patriots had to resist and push back against the Native Americans who, in a quasi-alliance with Great Britain, were pursuing their own objective of stopping incursions into their land. Third, the Revolution was in every sense a civil war. The Patriots had to politically and militarily defeat the Loyalists among the population of the American colonies. Finally, the Patriots had to deal with their own ranks. Within the group that we might label "Patriots" were a number of factions held together in a shaky confederation that was constantly threatened by sectional and cultural cleavages. One key reason that the Patriots eventually achieved their goals was that they succeeded in maintaining just enough unity to outlast their enemies. Once the war was won, the political and cultural differences among the Patriots grew and gave birth to political parties that nearly went to war with each other in the late eighteenth and early nineteenth centuries. Ultimately, the factional divides within the citizenry of the new nation devolved into an extraordinarily bloody civil war. The miracle of the Patriot insurgency is that they could delay the sectional fracture long enough to win and sustain nationhood.

Another unique aspect of this study is its greater attention to Patriot naval activities rather than the land campaigns. Without doubt, the operations of the Continental Army and Patriot auxiliary and irregular forces were crucial to success, but we have chosen to summarize rather than analyze the land campaigns for two reasons. First, there is already a wealth of historical material on the campaigns and battles of the American Revolution. Second, the objective of achieving nationhood cast the relationship between land and naval campaigns into a new light. At best, the land campaigns could stave off defeat, keep the British from sustaining political control over American territory, and bleed the British army to increase the cost of the war. However, to actually achieve independence, the Patriots had to bring decisive pressure on Parliament. Because the House of Commons and House of Lords comprised men with deep commercial connections extending throughout Europe, the Caribbean, North America, and the Indo-Pacific region, the best way to exert pressure would be at sea. When the East India Company, other trading interests, and English merchants saw their profits disappearing because of the extended conflict and the depredations of American and French privateers, they in turn brought pressure on the government to end the war. Hence, in our view, the land campaigns provided critically important defense, but the naval campaigns, in conjunction with diplomacy and non-importation movements, resulted in decision.

The Patriot insurgency offers powerful lessons in how to integrate the components of successful irregular warfare: ideology, political networking, communications, financial organization, logistics, military training, and a host of others. The student of modern warfare can learn much from understanding the worries and triumphs of George Washington, Nathanael Greene, John Adams, Thomas Jefferson, and others.

ORGANIZATION OF THE STUDY

ARIS Tier 1 Insurgency Case Studies are organized into five major sections:
- Introduction and Summary
- Context and Catalysts of the Insurgency
- Structure and Dynamics of the Insurgency
- Government Countermeasures
- Conclusion

This *Introduction and Summary* section presents an introduction to the ARIS series and a description of how the content in each particular case is presented. This section also includes a discussion of the types of sources and methods that were used to gather and analyze the data, as well as any methodological limitations encountered in the research. Last, this section includes a synopsis of the case study on the Patriot insurgency.

The section on *Context and Catalysts of the Insurgency* is divided into four chapters that address various aspects of the context within which the insurgency occurs. This section looks at the following elements:
- Physical environment
- Historical context
- Socioeconomic conditions
- Government and politics

The authors decided to discuss the organization and actions of the Patriot insurgency beginning in 1763 following the British government's Proclamation of 1763, which aimed at restricting the colonists' access to trans-Appalachia. Thus, everything that helped to shape the Patriot insurgency before 1763 will be summarized in the Historical Context chapter.

The organization and inner workings of each of the Patriot insurgency are analyzed in the *Structure and Dynamics of the Insurgency* section. This analysis considers various characteristics, including the following:
- Leadership and organization
- Ideology
- Legitimacy
- Motivation and behavior

- Operations
- External actors and transnational influences
- Finances, logistics, and sustainment

The *Government Countermeasures* chapter examines the political, military, informational, and/or economic actions taken by the British government and by external forces in support of the government to counter the efforts of the insurgency. This chapter is presented chronologically, broken down by separate political administrations or by significant counterinsurgency campaigns or initiatives.

The final chapter, *Conclusion*, provides observations about the aftermath of the revolution, considering questions, such as did any of the revolutionary factions succeed in changing any political, economic, or social conditions as attempted? What changes took place over the time frame of the study—to the government itself as well as to the movement (e.g., did the insurgent group disappear, become the ruling government, become a legitimate political party, etc.)? This chapter includes a discussion about which objectives or goals of the opposing sides were met and which were not and what compromises or concessions, if any, were made by either side.

METHODOLOGY OF THE STUDY

All ARIS Tier 1 Insurgency Case Studies are presented using the same framework. While not a strict template, it is a method used by the team to ensure a common treatment of the cases, which will aid readers in comparing one case with another.

All of the sources used in preparation of these case studies are unclassified and for the most part are secondary rather than primary sources. Where we could, we used primary sources to describe the objectives of the revolution and to give a sense of the perspective of the revolutionary or another participant or observer. This limitation to unclassified sources allows a much wider distribution of the case studies while hindering the inclusion of revealing or perhaps more accurate information. We selected sources that provide the most reliable and accurate research we could obtain, endeavoring to use sources we believe to be authoritative and unbiased.

These case studies are intended to be strictly neutral in terms of bias toward the revolution or those to whom the revolution was or is directed. We sought to balance any interpretive bias in our sources and in the presentation of information so that the case may be studied without any indication by the author of moral, ethical, or other judgment.

While we used a multi-methodological approach in our analysis, the analytical method that underpins these case studies can most accurately be described as contextual social/political analysis. Research in the social sciences is often done from one of two opposing perspectives. The first is a positivist perspective, which looks for universal laws to describe actions in the human domain and considers context to be background noise. The second is a constructivist perspective, which focuses almost entirely on the local context at the exclusion of any understanding of social or political structures or processes. Contextual analysis is "something in between."[5] It balances these two perspectives, combining an understanding of the actors, events, activities, relationships, and interactions associated with the case of interest with an appreciation for the significant role context played in how and why events transpired.

The term *context* is often used interchangeably with *environment*; however, the concept of context as used in these studies is much broader. Context includes factors, settings, or circumstances that in some way may act on or interact with actors, organizations, or other entities within the country being studied, often enabling or constraining actions. It is a construct or interpretation of the properties of a system, organization, or situation that are necessary to provide meaning above and beyond what is objectively observable.[6]

Although we have applied this methodology throughout these case studies, the section entitled *Context and Catalysts of the Insurgency* focuses very heavily on contextual aspects. Examples of elements of context often used in this type of analysis include culture, history, place (location), population (demography), and technology. Within these studies, we present the primary discussion of context as follows.

Physical Environment

Social scientists often cite features of the physical environment as a risk factor for conflict—whether it is slope elevation, mountainous terrain, or rural countryside. Rough terrain[7] is a typical topographical feature correlated with rebel activity, as it provides safe havens and resources for insurgents. Insurgent groups such as the Afghan Taliban benefited from the mountainous terrain of Afghanistan, making pursuit and surveillance by countervailing forces difficult. Likewise, the Viet Cong in Vietnam benefited from dense forest cover despite American attempts at defoliation.[8] In the case of the American Revolution, Patriot leaders often made great use of rough terrain. Francis Marion eluded capture by British Colonel Banastre Tarlton during operations in South Carolina. Likewise, the Patriots used the rough terrain of the Lake Champlain-Hudson Valley corridor to slow and eventually defeat British General John Burgoyne.

Less clear are the reasons behind the correlation that researchers have found between rough terrain and conflict. Most theories for this relationship center on insurgent viability and a state's capacity to govern. In short, rough terrain is correlated with conflict, but that does not mean it causes conflict or that rough terrain is necessary for a conflict to emerge.[9, 10]

Researchers have argued that other geographic features, such as location and distance, impact conflict patterns and processes. Generally, research has shown that regions farther from the capital are at higher risk for conflict, as are those closer to international borders. Another important consideration when analyzing the impact of geography on conflict patterns and processes is the expanse of the conflict. While it is common to speak of entire countries embroiled in conflict, studies show that the actual conflicts generally occur only in a small percentage of a state's territory, typically 15 percent. Despite that low figure, however, recent empirical work has shown that internal conflicts can sometimes encompass nearly half of the territory of the host country.[11]

Historical Context

Revolutions or insurgencies do not emerge from formless ether, but rather take their shape from accumulated layers of historical experience. Not only are actors in insurgent movements important participants in history, but they are also its end users. That is, insurgent movements are not only shaped by historical experience, but they also seek to understand and manipulate the key components—whether historical events, persons, or narratives—of those experiences to accomplish their objectives. Thus, sustained, organized political violence cannot be adequately explained without analyzing the historical context in which it developed. Some of the themes analyzed in this section are the legacies, whether organizational, political, or social, of conflict over time; the formation of group and organizational identity and its attendant narrative; the development of societal and political institutions; and the changing relationships, and perceptions thereof, that balance national, local, and/or group interests.[12]

Charles Tilly[13] made the following observations about the relationship between social movements and historical context:

- "Social movements incorporate locally available cultural materials such as language, social categories, and widely shared beliefs; they therefore vary as a function of historically determined local cultural accumulations.
- Social movements occurring in adjacent places such as neighboring countries influence local social movements, hence, historically variable adjacencies alter the kinds of social movement that appear in any particular place.
- Path dependency prevails in social movements as in other political processes, such that events occurring at one stage in a sequence constrain the range of events that is possible at later stages.
- Once social movements had occurred and acquired names, both the name and competing representations of social movement became available as signals, models, threats, and/or aspirations for later actors."[14]

While Tilly's observations address social movements usually understood to be nonviolent political movements, he and his collaborators argued that contentious political activity belonged on a continuum, not in separate categories.[15] Violent and nonviolent groups belonged to the same genus but used different "repertoires of contention."[16] Thus, the same methodologies used to explain nonviolent political activity could also be useful in explaining violent political activity. Our extensive research on nearly thirty insurgencies supports this theory. The insurgencies, but also the individual participants themselves, often began their careers by engaging in nonviolent political activity, transitioning to violence sometimes only after many years. To connect these more explicitly with revolutionary and insurgent activities, we examine each of these general observations of social movements and apply them to the specific activities associated with an insurgency or revolution. Revolutions and insurgencies typically begin as local or regional movements, and as such, they include all of the aspects of local cultural material, which, as previously mentioned, contributes to the ontology of a social movement.

Activities associated with insurgency frequently cross borders and have an influence on the societies and movements in adjacent regions. Actions taken by an insurgent organization at one point in time can eliminate or enable possible future options for furthering the insurgency. Groups associated with revolutions and insurgencies usually seek recognition for their actions, so it is important for them to have names and symbols (emblems, flags, etc.) that can be easily associated with them and their causes. These representations then become the public branding of the organization and are used by supporters and detractors alike to further the narrative or counternarrative. Given these factors, the historical context within which any insurgency, revolution, or other internal conflict occurs is a critical element in analyzing and understanding these phenomena.

Socioeconomic Conditions

The relationship between high per-capita (or per-person) gross domestic product (GDP) and political stability is among the most robust findings of social scientists studying conflict dynamics. Conflict research can be divided roughly into three categories: onset, duration, and termination. Some of the relevant socioeconomic theories for domestic political violence, where the latter range from riots and rebellion to civil war, include economic deprivation; political, social, and economic grievances; ethnic nationalism; opportunity costs or greed; and the resource curse.

Some political scientists argue that countries with lower levels of economic development are more likely to witness political violence,[17] further suggesting that a country's low GDP per capita is a proxy measure for poor state capacity, where the central state government has limited ability to project its power throughout its territory.[18] Although low economic development has been demonstrated to be a risk factor for rebellion, "poverty is generally considered to be an indirect contributing factor and not a primary cause of political violence."[19] Another perspective argues that poverty and the attendant lack of political, social, and economic modernity themselves do not generate political instability, but rather efforts to achieve modernity in these realms lead to political instability.[20] Hence, the modernization process is inherently conflictual, as elite members of the ancient regime may see their fortunes decline while new social classes become empowered within a political system incapable of articulating and responding to their unique demands.

Yet another perspective is that poverty reduces opportunity costs (greed) for rebels, making recruitment and other functions easier and more attractive.[21] Thus, poverty provides a possible indicator for where insurgency is likely, as it generally also describes poor governance and corruption, which enable insurgents to gather and operate. Other poverty-related indicators that may lead to rebellion include rural poverty and greater landless populations,[22] selective benefits,[23] and the probability of greater gains through rebellion.[24] However, poverty itself is not enough to predict insurgency.

Another prominent socioeconomic reason for conflict is relative deprivation. In *Why Men Rebel*, Gurr argues that political violence can be explained by relative deprivation, or the sense among one segment of the population that it is entitled to the goods and services that they themselves are not able to attain but to which others in the population have access.[25] If political allegiance is based on ethnicity and one ethnic minority group experiences deprivation relative to the ethnic majority group (as happened with the Tamils in Sri Lanka vis-à-vis the Sinhalese in the early 1970s), then the minority may give up hope for satisfying its aspirations within the unitary state and seek to detach itself from the body politic of the nation.[26]

Other important indicators for grievance are political exclusion and economic inequality. In Colombia, for example, the political vacuum after *La Violencia* and the inadequate national compromise that followed disenfranchised several groups, especially communist and socialist groups. This reinforced Colombia's historical inability to include all its citizens in its political process, leading to political exclusion and the economic space and motivation for insurgency by both political and criminal groups. However, like economic deprivation, relative deprivation is better understood as a general risk factor for conflict rather than as a predicting cause.[27] These are often indirect causes for political violence and are better understood as risk factors for violence.

Unlike the grievance argument, another proposed socioeconomic factor associated with insurgencies is low opportunity cost, or greed, in which groups are motivated by economic reasons for pursuing conflict.[28] From this perspective, the occurrence of conflict is due to opportunity or viability—these are the conditions "sufficient for profit-seeking, or not-for-profit, rebel organizations to exist."[29] Another part of opportunity costs is natural resources, which allow rebels to profit from looting, adding to the greed motives that the presence of lootable goods will increase the chance of rebellion.[30]

Ethnic wars, while influenced by both economic deprivation and relative deprivation, are driven by identity. In the case of ethnic conflict, "poverty and low levels of economic development could increase ethnic conflict," but there is a "weaker relationship between poverty/underdevelopment and ethnic war." In the case of ethnic war, the driving issue is that of identity and is not merely economic.[31]

More specifically, the clash of ethnic identities and fears of cultural extinction often provide the animus motivating ethnic conflicts. Benedict Anderson[32] defined a nation as "an imagined political community . . . it is imagined because the members of even the smallest nation will never know most of their fellow-members, meet them, or even hear of them, yet in the minds of each lives the image of their communion."[33] The same as well can be said of ethnic groups, and as in the case of Tamils and Sinhalese in Sri Lanka, ethnic categories and the symbols and meanings ascribed to them often embody social constructions derived from both facts and distortions of the historical record. Thus, while separate south Indian and Sinhalese communities have resided on the island for several thousand years, during the recent conflict, some participants may have "read history backwards"[34] and viewed past conflicts through the prism of an ethnic conflict paradigm, irrespective of whether the participants of the conflicts in the distant past were motivated by ethnic grievances.

Government and Politics

When considering politics in the contextual analysis of insurgency and revolution, the focus is on understanding the influence and impact of ideas on individuals and groups on both sides of the conflict and understanding how those ideas shaped the decisions and actions of the various actors. The contextual aspects of government suggest the need for understanding the national and local political discourse, the resolution of competing political objectives, and government institutions that are in place and their impact to the ability of an insurgent group to operate and achieve its goals.

We discussed insurgency or revolution as a specific instance of a social movement. Social movements have been defined as "networks of informal interactions between a plurality of individuals, groups, or associations, engaged in a political or cultural conflict, on the basis of a shared collective identity."[35] Government and politics is one of the primary means through which ideas are enacted within society. Social movements are another example where organizations attempt to achieve social outcomes. The key difference between social movements and other means within society is that social movements (1) exhibit strong lines of conflict with political or social opponents, (2) involve dense interorganizational networks, and (3) are made up of individuals whose sense of collective identity exists beyond any specific campaign or engagement.[36]

There are several different schools of thought on the political factors associated with revolution and insurgency. Many researchers stress the importance of the economic resources available to both the regime and insurgents, but this discussion has several variants. One theory that considers economic factors as critical indicators of intrastate conflict includes the financial opportunities for rebels, as well as the economic capacity of the regime to provide public services (e.g., defense) as key parameters.[37] Other researchers focus on the ability of either side to finance military operations, the ability of insurgents to capitalize on an unstable political environment, and control over natural resources for financial gain as indicators for conflict.

More recently, social scientists shifted their focus to look at regime type, political structures, and relationships as the primary indicators, rather than simply economic resources or opportunities. Simply put:

> Most states have potential insurgents with grievances and resources, but almost always possess far greater military power than do insurgents. A united and administratively competent regime can defeat any

insurgency; it is where regimes are paralyzed or undermined by elite divisions and state-elite conflicts that revolutionary wars can be sustained and states lose out to insurgencies.[38]

The assertion is that the type of government that is in place within the country is one of the most significant factors. Many of the initial studies on this topic used a simple categorization of regimes as either democratic or autocratic, but researchers have also adopted a three-way categorization that includes democracy and autocracy as categories, as well as a middle category of "anocracy," which characterizes a polity that has both democratic and autocratic elements and which may find itself undertaking the risky transition from autocracy to democracy. From the three-category perspective, an anocracy is more likely to experience violence than a democratic or autocratic regime. In an anocracy, power or control does not rest solely with government institutions but rather is vested in multiple competing groups that continually struggle against one another and the government for control. Some researchers, however, considered these simple categorizations to be overly simplistic or ambiguous. Recent work develops a more detailed set of parameters to determine what the researchers call "the institutional character of the national political regime." After considerable research, experts found this attribute to be the most significant indicator or predictor of conflict.[39]

SYNOPSIS OF CASE STUDY IN INSURGENCY AND REVOLUTIONARY WARFARE: PATRIOT INSURGENCY (1763–1789)

Timeline

1754–1763	Great Britain and the English colonies fight and eventually win against French Canada and their Indian allies in the French and Indian War. The debt from the war leads to Britain's taxation of the colonies—a major cause of the Patriot insurgency.
1757	William Pitt the Elder becomes Prime Minister of Britain. His leadership turns the war in Britain's favor.

1758	The Royal Navy and British Army seize Louisbourg from the French. The British also capture Fort Frontenac and Fort Duquesne (later "Pittsburgh").
1759	"Annus Mirablis." The Royal Navy defeats the French Navy at Quiberon Bay, preventing French reinforcement of Canada. British General James Wolfe defeats Montcalm at Quebec.
1760	French Canada capitulates to the British.
October 25, 1760	George III ascends the throne of Great Britain.
February 10, 1763	Treaty of Paris concludes the war, awarding all of North America east of the Mississippi (including Canada) to the British Empire.
October 7, 1763	Proclamation of 1763 seeks to prevent American colonists from migrating over the Appalachian Mountains.
April 5, 1764	Parliament passes the Sugar Act that aims at preventing American colonists from importing foreign sugar products, including molasses and rum.
September 1, 1764	Parliament passes the Currency Act that requires American colonists to pay British creditors in hard currency, rather than in printed money.
March 22, 1765	Parliament passes the Stamp Act, setting off a crisis in the American colonies. Patriots are outraged, insisting that the Act is a form of direct taxation that is unconstitutional.
March 24, 1765	Parliament passes the Quartering Act, imposing the burden of housing British regulars on American colonists, further inflaming the embryonic insurgency.
May 30, 1765	Virginia passes the Virginia Stamp Act Resolutions, defying Parliament and insisting that Virginia colonists are British citizens entitled to the right to tax themselves.
October 1765	The Stamp Act Congress resolves that only the people themselves may, through their representatives in colonial assemblies, pass tax legislation.
March 18, 1766	Parliament withdraws the Stamp Tax but passes the Declaratory Act, reiterating that Parliament has the right to legislate for the colonies and that the colonies are subordinate to the mother country.

June 29, 1767	Parliament passes the Townshend Revenue Act, imposing taxes on lead, paint, tea, and other commodities. This touched off another round of protests by the Patriots. British troops occupied Boston to put down the revolt, whereupon Boston, and later all the colonies, adopted nonimportation resolutions that dried up British commerce in America.
August 1, 1768	Boston merchants and traders band together and agree to impose the Boston Non-Importation Agreement.
March 5, 1770	British troops shot into harassing citizens, killing several, in an incident that Patriots called the "Boston Massacre."
June 1772	The Gaspee Affair. Colonists attacks and destroys a grounded British revenue collecting ship. Parliament is outraged and demands the arrest of the perpetrators, but no colonial authorities cooperate.
May 10, 1773	Parliament passes the Tea Act, an attempt to cajole colonists into buying tea from the East India Company with lower prices to undercut higher priced smuggled tea. However, because the cheaper tea is still subject to the Townshend duties, colonists resist the measure.
December 16, 1773	The Boston Tea Party. Patriots in Boston boards British ships and throws tea into the harbor, inflicting serious economic loss on the owners.
March-June 1774	The Coercive Acts, in response to the Tea Party and other acts of rebellion, punishes Boston by closing its port until the inhabitants paid for the destroyed tea, removes colonial control of the administration of justice in some cases, limits the rights of Massachusetts to self-government, imposes more quartering, and gives preferential treatment to Quebec.
September-October 1774	First Continental Congress meets and issues Resolves that reiterate American rights and requests repeal of the Intolerable Acts.
October 20, 1774	The Articles of Association calls for universal non-importation and nonconsumption of British goods and establishes citizens' committees to enforce the measures.

April 19, 1775	Patriot Minutemen and British regulars clash at Lexington and Concord. The battle becomes known as "the shot heard 'round the world."
June 15, 1775	The Second Continental Congress names George Washington as Commander-in-Chief of the Continental Army.
June 17, 1775	Outside of Boston, British regulars drive Patriot militia from Breed's Hill, but at the cost of over a thousand killed or wounded. Known as the Battle of Bunker Hill.
November 7, 1775	The Royal Governor of Virginia, Lord Dunmore, issues a proclamation offering freedom to slaves and indentured servants who would run away and enlist to fight against the Patriots.
December 1775	Patriot forces invade Canada but suffer defeat before the walls of Quebec.
January 10, 1776	Thomas Paine publishes *Common Sense* anonymously to promote independence and republican government.
March 17, 1776	The British evacuate Boston and relocate to Halifax, Nova Scotia. Patriots view this as a great victory.
June 28, 1776	Patriot forces repel a British naval attack on Charleston, South Carolina.
July 4, 1776	The Continental Congress signs the Declaration of Independence.
August 17, 1776	British regulars defeat George Washington's Continental Army at the Battle of Long Island. British forces occupy Manhattan.
November 1776	The British, including Hessian mercenaries, capture Forts Washington and Lee, on the Hudson River, precipitating a long, difficult retreat across New Jersey by the Continental Army. Washington's troops—demoralized and worn out—escape across the Delaware River.
December 26, 1776	Washington leads the Continental Army across the Delaware River and conducts a surprise attack on Trenton, capturing Hessian troops and needed supplies. This unforeseen victory bolsters Patriot morale.

January 3, 1777	Washington scores a second surprise victory at Princeton, New Jersey.
September 1777	British forces invade Pennsylvania, defeat Washington and occupy the Patriot capital at Philadelphia. Washington withdraws to Valley Forge, where his army suffers greatly during the winter.
October 17, 1777	After invading northern New York from Canada, the British General Burgoyne surrenders to Patriots at Saratoga, bringing his ill-fated campaign down the Hudson Valley to a close. This Patriot victory leads to the French decision to intervene and join the Patriots.
February 6, 1778	France and the United States sign the Treaty of Alliance, which brings the French into the American War for Independence.
June 28, 1778	After withdrawing from Philadelphia, British forces fend off pursuit by Washington's army at Monmouth, New Jersey. Patriot soldiers display improved discipline after a winter of hard training at Valley Forge.
1779–1781	British open a campaign in the southern colonies, capturing Savannah. This move initiates a brutal war combining conventional and irregular forces.
March 1, 1781	The thirteen Patriot states adopt their first constitution, the Articles of Confederation. This weak government struggles for lack of its own revenue and control over trade.
October 19, 1781	British General Cornwallis, surrounded on land and sea by Patriot and French forces, surrenders in the last major battle of the war within the United States.
April 1782	In the Battle of the Saintes, in the Caribbean, a British fleet crushes a French fleet, saving the British colonies in the West Indies
September 3, 1783	The United States and Great Britain sign the Treaty of Paris, ending the Revolutionary War. In November, British troops depart New York City. Washington resigns as Commander-in-Chief in December.
September 17, 1787	A new US Constitution, establishing a stronger national union, is signed by most of the delegates who attended a convention in Philadelphia.

June 21, 1788	The US Constitution is adopted, when New Hampshire becomes the ninth state to ratify.
1788-1794	War persists in the Ohio country, where Indian nations confederate to block American expansion. The British covertly supply the natives with arms and ammunition.

ENDNOTES

[1] The terms *insurgency* and *revolution* or *revolutionary warfare* are used interchangeably in the ARIS series. We adopted the term *revolution* to maintain consistency with the Special Operations Research Office (SORO) studies conducted during the 1960s, which also used the term. Many social scientists use an arbitrary threshold of battle deaths to delineate civil war from other acts of armed violence. Our definition relied on Charles Tilly and Sidney Tarrow's definition of contentious politics, activity that "involves interactions in which actors make claims bearing on someone else's interests or programs, in which governments are involved as targets, initiators of claims, or third parties."

[2] Charles Tilly and Sidney Tarrow, *Contentious Politics* (Boulder, CO: Paradigm Publishers, 2007), 4.

[3] Chuck Crossett, ed. *Casebook on Insurgency and Revolutionary Warfare Volume II: 1962–2009* (Fort Bragg, NC: United States Army Special Operations Command, 2012), xvi.

[4] Ibid., xii–xiii.

[5] Charles Tilly and Robert E. Gordon, "It Depends," in *The Oxford Handbook of Contextual Political Analysis*, eds. Robert E. Gordon and Charles Tilly (Oxford: Oxford University Press, 2006), 9.

[6] W. B. Max Crownover, "Complex System Contextual Framework (CSCF): A Grounded-Theory Construction for the Articulation of System Context in Addressing Complex Systems Problems" (PhD diss., Old Dominion University, 2005).

[7] Most researchers use mountains (or slope elevation) and forests as a proxy for "rough terrain." Little attention has been paid to other topographical features, such as swamps, that impede government access or surveillance.

[8] Nathan Bos, "Underlying Causes of Violence," in *Human Factors Considerations of Undergrounds in Insurgencies*, ed. Nathan Bos (Fort Bragg, NC: United States Army Special Operations Command, 2013), 27.

[9] The relationship between terrain and conflict can be described as follows: "rebels who seek refuge in the mountains are better able to withstand a militarily superior opposition . . . that rebel groups will take advantage of such terrain, whenever available. We do not believe that terrain in and of itself is a cause of conflict, nor does the rough terrain proposition anticipate such a relationship."

[10] Halvard Buhaug and Jan Ketil Rød, "Local Determinants of African Civil Wars, 1970–2001," *Political Geography* 25, no. 3 (2006): 316.

[11] Clionadh Raleigh, Andrew Linke, Håvard Hegre, and Joakim Karlsen, "Introducing ACLED: An Armed Conflict Location and Event Dataset: Special Data Feature," *Journal of Peace Research* 47, no. 5 (2010): 652.

[12] Charles Tilly, "Why and How History Matters," in *The Oxford Handbook of Contextual Political Analysis*, ed. Robert E. Gordon and Charles Tilly (Oxford: Oxford University Press, 2006), 423.

[13] Charles Tilly is widely regarded as one of the twentieth century's most important theorists of the social dynamics of political conflict.

[14] Tilly, "Why and How History Matters," 425.

[15] Doug McAdam, Sidney G. Tarrow, and Charles Tilly, *The Dynamics of Contention* (Cambridge, UK: Cambridge University Press, 2001).

[16] Ibid.

[17] Bos, "Underlying Causes of Violence," 15.

18. James D. Fearon and David D. Laitin, "Ethnicity, Insurgency, and Civil War," *American Political Science Review* 97, no. 1 (2003): 75–90.
19. Bos, "Underlying Causes of Violence," 15.
20. Samuel P. Huntington, *Political Order in Changing Societies* (New Haven: Yale University Press), 1968.
21. Paul Collier and Anke Hoeffler, "Greed and Grievance in Civil War," *Oxford Economic Papers* 56, no. 4 (2004): 563.
22. Jeffery Paige, *Agrarian Revolution* (New York: The Free Press, 1975).
23. Mark Lichbach, *The Rebel's Dilemma* (Ann Arbor: University of Michigan Press, 1995).
24. Madhav Joshi and David Mason, "Between Democracy and Revolution: Peasant Support for Insurgency versus Democracy in Nepal," *Journal of Peace Research* 45, no. 6 (2008): 765–782. See also David Mason, "Land Reform and the Breakdown of Clientelist Politics in El Salvador," *Comparative Political Studies* 8, no. 4 (1986): 487–517.
25. Bos, "Underlying Causes of Violence," 17.
26. Ted Robert Gurr, *Why Men Rebel* (Princeton: Princeton University Press, 1970), 571.
27. Bos, "Underlying Causes of Violence," 17.
28. Collier and Hoeffler, "Greed and Grievance in Civil War," 565.
29. Ibid.
30. Ibid., 564.
31. Nicholas Sambanis, "Do Ethnic and Nonethnic Civil Wars Have the Same Causes?: A Theoretical and Empirical Inquiry (Part 1)," *Journal of Conflict Resolution* 45, no. 3 (2001): 259, 266.
32. Benedict Anderson, *Imagined Communities* (London: Verso, 2006), 6.
33. Ibid.
34. Chelvadurai Manogaran and Bryan Pfaffenberger, "Introduction: The Sri Lankan Tamils," in *The Sri Lankan Tamils: Ethnicity and Identity* (Boulder, CO: Westview Press, 1994), 20.
35. Mario Diani, "The Concept of Social Movement," *Sociological Review* 40, no. 1 (1992): 1–25.
36. Mario Diani and Ivano Bison, "Organizations, Coalitions, and Movements," *Theory and Society* 33 (2004): 281–309.
37. Collier and Hoeffler, "Greed and Grievance in Civil Wars," 563–595.
38. Jack A. Goldstone, Robert H. Bates, David L. Epstein, Ted Robert Gurr, Michael B. Lustik, Monty G. Marshall, Jay Ulfelder, and Mark Woodward, "A Global Model for Forecasting Political Instability," *American Journal of Political Science* 54, no. 1 (2010):190–208.
39. Ibid.

PART I.
CONTEXT AND CATALYSTS OF THE INSURGENCY

CHAPTER 2.
PHYSICAL ENVIRONMENT

America is a land of wonders, in which everything is in constant motion and every change seems an improvement.

—Alexis de Tocqueville

Chapter 2. Physical Environment

THE GEOGRAPHY OF A REVOLUTION

The American Revolution unfolded within a widening sphere that centered on New England and the city of Boston. At the start of the rebellion in 1775, British authorities believed the crisis was a strictly local problem and that by exerting military power in New England, the uprising could be contained and defeated.

The Patriot insurgency disproved this concept the following year with the signing of the Declaration of Indepenence by delegates from all thirteen colonies (see Figure 2-1). The sphere of rebellion thus widened to include the entire Atlantic seaboard—potentially involving British Canada as well.

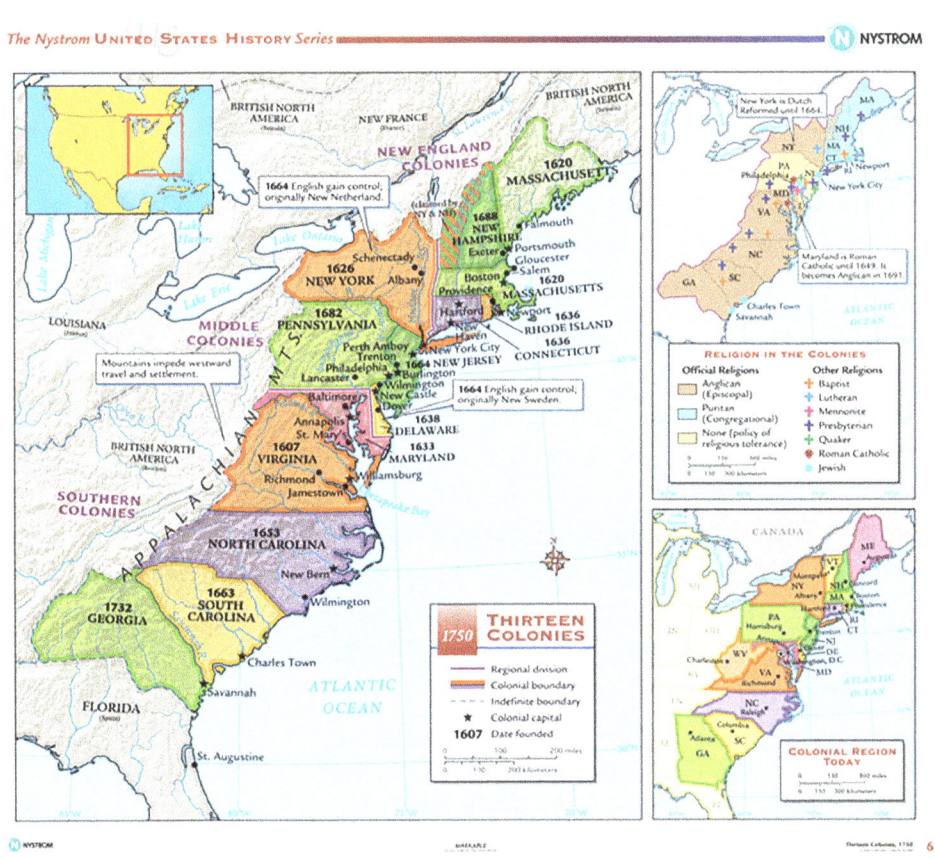

Figure 2-1: The thirteen colonies.

Finally, in 1778, French forces joined the Patriots against the British, and the conflict became global. British and French interests in the Caribbean framed the war in the colonies as the Royal Navy and French Navy vied for control of the economically vital sugar islands.[1] Spain entered the war as an ally of France the following year, further distracting British forces, although they did not directly assist the Patriots.

Time and Distance

One of the key factors that determined the course of the Patriot insurgency was the distance between the mother country of England and the colonies. Over three thousand miles separated colonial America from London, and the trip by sea took one-to-two months eastbound, while the westbound journey took longer. The separation underscored the loose political connections between England and her colonial possessions.

The long distances also complicated military planning and communications, most famously during the British campaigns of 1777 that ended in disaster at Saratoga. The Patriots likewise had a difficult time coordinating with their French allies and especially the naval forces.

Infrastructure

Revolutionary America had very poor infrastructure beyond the six major ports of Boston, New York, Philadelphia, Norfolk, Charles Town, and Savannah. Roads were few and of poor quality. The paucity of road networks made the projection of British combat power difficult and allowed the Patriots to preserve their armies at a safe distance when they wanted to. Navigable waterways offered the British opportunities to push both floating firepower and British regulars into key positions, most notably during the 1776 battles around Long Island. The Patriot leaders quickly learned that exposing their armies to envelopment near the coast was a foolhardy endeavor. Conversely, the British could scarcely hope to find, pin, and finish off Patriot forces deep in the hinterland.

The thick forests that covered much of America made conventional military operations difficult. Adapting to frontier conditions required British generals to make use of their Indian allies to navigate, scout, and raid through the rough terrain. Likewise, the rough interior of the southern colonies influenced leaders on both sides of the conflict to form fast-moving irregular bands that could hunt down and kill or capture Loyalist or Patriot insurgents.

Key Regions

Ports

The major ports that influenced the course of the Patriot insurgency were Quebec, Halifax, Boston, New York City, Philadelphia, Norfolk, Charles Town, and Savannah. In addition to these, there were numerous smaller ports, especially in New England and the Middle Colonies. Ports were critical locations during the insurgency because (1) they were closely regulated as an expression of British trade policy; (2) they were easily accessible to the Royal Navy; (3) they were locations of great economic importance for trade, communications, and employment; and (4) they presented opportunities for Patriots and others to smuggle goods into and out of the country.

Until the French intervention in 1778, the Royal Navy enabled the British to project combat power easily into any port and from there into the nearby countryside. Consequently, it proved very dangerous for the Patriots to attempt to defend such ports, as Washington discovered in 1776 during his failed attempt to defend New York City. After the French intervention, control of the ports became contested through the rest of the war. The French could now project combat power through the ports and linkup with Patriot armies, but the coordination of such moves continued to be a challenge until the summer of 1781.

Hudson Valley-Lake Champlain Coridor and Canada

The Hudson Valley-Lake Champlain corridor (see Figure 2-2) connected the colony of New York with Canada—formerly French but recently conquered by the British. During the Revolutionary War, both the British and Patriots conducted military campaigns along this avenue, and both sides suffered the logistical difficulties associated with trying to move armies through rough country with little infrastructure. Because large armies could traverse the region only along waterways, adversaries built fortifications at chokepoints to block them.

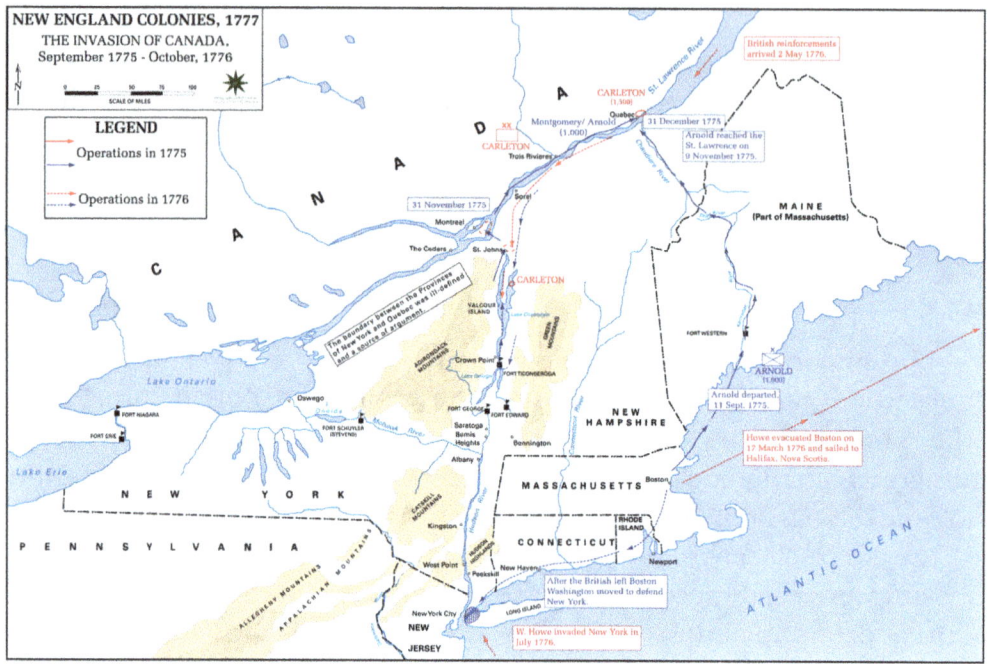

Figure 2-2: The Hudson Valley-Lake Champlain corridor.

The Patriots used the corridor to invade British Canada in the summer of 1775. They also conducted a supporting attack down the Chaudierre River. This initial effort by elements of the Continental Army faltered at the Battle of Quebec in December 1775, but the campaign underscored for both sides the military importance of the Lake Champlain corridor.

The corridor led to the St. Lawrence River valley and the key cities of Montreal and Quebec. It likewise allowed the British to project combat operations into the Middle Colonies and thereby cut off the western flank of New England. This idea lay behind General John Burgoyne's ill-fated campaign of 1777 that led to the British disaster at Saratoga.

Mohawk Valley and the Great Lakes

Connected to the Hudson Valley at almost a right angle was the Mohawk Valley, connecting the Great Lakes to the interior of New York. This avenue of approach remained a concern for both sides as they attempted to dominate the Hudson Valley because a military force operating along the Mohawk River could outflank forces based in Albany.

New England

Massachusetts (including present day Maine), New Hampshire, Connecticut, and Rhode Island comprised New England. The Patriot insurgency was born in New England, and the area and the leaders who emerged from it were among the most important political forces during the struggle. The intense economic conflict between Britain and the American colonies had its epicenter in New England and its main city, Boston.

The focus of military operations, however, migrated away from New England early in the war. The British occupation of Boston, 1775-1776, had run its course and accomplished nothing other than further alienating the Americans from London. When George Washington's embryonic Continental Army maneuvered to threaten British forces with bombardment, British General Howe evacuated the city and did not return. Instead, the rest of the war played out to the west and south of New England as the British continued to search in vain for a geographical center of gravity of the rebellion.

Middle Colonies

Pennsylvania, New York, New Jersey, Maryland, and Delaware comprised the Middle Colonies and were a major theater of war throughout the conflict. The two key ports of New York City and Philadelphia allowed the British to push combat power into the region, and they correctly anticipated that they would find more Loyalist support there than in New England. The Patriots—primarily the Continental Army and its associated detachments and militias—operated from 1776 through 1778 along the western fringe of British-controlled territory, frustrating the British leadership as they attempted to bring the Patriots to a decisive, final battle.

The Middle Colonies provided Washington logistical support and defensible bases once he realized that he could not defend the ports or coastline against an enemy that dominated the seacoast. From the Middle Colonies, he could send military detachments north through the Hudson Valley to threaten Canada, or south into the Southern Colonies (see Figure 2-3).

Case Studies in Insurgency and Revolutionary Warfare—The Patriot Insurgency

Southern Colonies

Virginia, North and South Carolina, and Georgia hosted the final, drawn-out campaigns that led to the end of the Revolutionary War. Frustrated in their attempts to quell the rebellion in New England and the Middle Colonies, the British opened a campaign in the south in December 1778 by seizing Savannah and took Charleston in May 1779. The rationale for the move was the hope that Loyalist forces would emerge and help the British army take decisive control of the vast agricultural region. They likewise intended to incite black slaves and Native Americans to join their cause. Instead, the southern campaign led to extremely bloody civil war between Loyalist and Patriot forces in the interior. As had occurred to the north, the Patriot armies operated just out of reach of the British. The battles that resulted continued to bleed both sides, but the British Parliament finally concluded that they could no longer afford the loss in blood and treasure, which led to the decision to end the conflict in 1783.

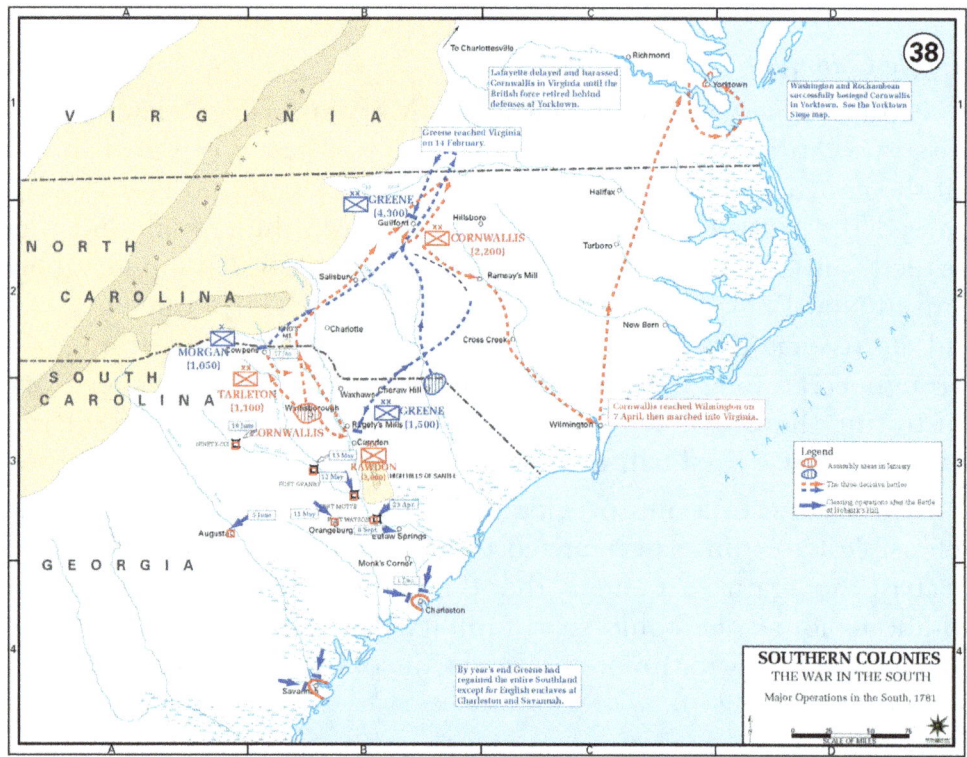

Figure 2-3: The southern colonies.

The West Indies

Although the Patriots deployed no combat power to the Caribbean, the West Indies were central to the conflict because both the British and the French possessed valuable sugar plantations in the islands. The British were particularly interested in defending Jamaica and their possessions in the Lesser Antilles. The French likewise wanted to protect their colonial holdings in the West Indies. Consequently, both the French and British navies continued to move forces back and forth from the Caribbean to the Atlantic seaboard throughout the conflict, often complicating joint coordination between armies and navies.

ENDNOTES

[1] Al M. Rocca, "The Impact of Geography on the American Revolution: Expanding Regions of British Military Responsibility," *Social Studies Review* 42, no. 2 (Spring 2003).

CHAPTER 3.
HISTORICAL CONTEXT

If there must be trouble, let it be in my day, that my child may have peace.

—Thomas Paine

INTRODUCTION

This chapter discusses three essential historical contexts that underlie the study of the Patriot insurgency from 1763 to 1789. First, there is a general discussion of the thirteen colonies that comprised the Atlantic seaboard that spawned the resistance against British rule. This is followed by an historical analysis of the Native American tribes that influenced the revolutionary struggle. The chapter concludes with an explanation of the French and Indian War of 1756-1763 that was the fundamental catalyst that led to the American Revolution.

THE AMERICAN COLONIES

The Patriot insurgency grew from the discontented populations of British North America. The thirteen colonies there sprang from diverse roots, boasted different economies, and featured a variety of religious affiliations and cultures. A brief enumeration of the colonies and their relationship with the British government follows.

South Carolina

Founded in 1670, South Carolina was initially part of the "Province of Carolina," granted to a collection of proprietors who sought to establish lucrative plantations. In 1719, pursuant to colonists rising against and overturning the proprietors and absentee landlords, it became a crown colony. Ten years later, North Carolina was split off into its own colony. South Carolina's main crops were rice, indigo, and tobacco. Planters who emigrated from Barbados also attempted to grow sugar and cotton there. The plantations required huge numbers of African slaves, and the colony became a main importer. South Carolina planters, whose power base was in the low country lived in constant fear of slave uprisings or any threat to their slave-based economy. The "Upcountry" had more Loyalists, many of whom feared attacks from the native tribes near them. As the Revolution approached, the colony threw in its lot with the Patriots, in part because the rebels would remain friendly to slavery while the British intended to incite the slaves to flee and join the Loyalist cause.

North Carolina

Sharing a similar history with South Carolina, the colony separated in 1729. Originally, the British population was engaged in subsistence farming, but with the influx of European immigrants in the eighteenth century, the economy expanded into crops and livestock that proved more lucrative. As the Revolution approached, the colony's Loyalists opposed the Patriot cause and participation in the First Continental Congress. However, their militia forces were defeated at the Battle of Moore's Creek in February 1776. Loyalists in the interest of maintaining both their liberty and their slaves, North Carolina joined the Patriot cause. Their contribution to the insurgency remained small-scale throughout the war, but they contributed some seven thousand soldiers to the Continental Army, along with local militia when needed.

Georgia

In the 1730s, James Oglethorpe led English settlers to what became the colony of Georgia in 1733, named after the sovereign, King George II. Although slavery was originally banned, in 1749, the colonial assembly overturned that ruling, and Georgia became a slave colony. As with the majority of other colonies, Georgian planters objected to Parliament's increasingly strict trade regulations and proposed taxes. The colony joined the Patriot cause, expelled their royal officials, and wrote a state constitution. The British occupied much of the colony through 1780.

Virginia

Founded in 1607 with the establishment of Jamestown (following earlier failed attempts), Virginia was a company colony through 1624, when it became a royal colony. The settlers teetered on the edge of extinction and failed to prosper until the introduction of tobacco farming during the 1610s. The Virginia planters at first employed indentured servants on the labor-intensive tobacco plantations increasingly resorted to black slavery. The colony grew to become the wealthiest and most populous of the thirteen colonies. The wealthy elites who controlled the colony through the elected General Assembly favored the Anglican Church and coveted status as aristocratic Britons. Their interests in acquiring land from the West, along with their aversion to the British taxation led the colony's leaders to join the Patriot cause.

Within the colonial legislature, the landed elite ruled with few divisions among them and, for the most part, in the interest of the general public of free, white men. Prior to the Revolution, there was little significant factionalism in Virginia, and the colony produced some of the Patriots' best and most resolute leaders.[1]

Maryland

Maryland was founded by royal charter by George Calvert in 1632. He died and passed the charter on to his son that same year. The colony of Maryland was known for its tolerance of Roman Catholics, and the first settlers who came there in 1633 were of that faith. Led by the Calvert family, the colony evolved a legislative assembly according to the English tradition. The ensuing conflict between Protestants and Catholics led to the 1649 Maryland Toleration Act—a trend away from state-sponsored religious homogeneity that the Patriots would eventually adopt. Maryland was at first resistant to the idea of independence from Britain, but the colony eventually joined the Patriot cause.

Pennsylvania

William Penn received a royal charter in 1681 and established Pennsylvania based on religious toleration as a proprietary colony. Penn's religious policy attracted immigrants of many different denominations, including Quakers, Catholics, Puritans, Calvinists, and others. Penn obtained proprietary rights to what would eventually become the state of Delaware as well. Under Penn and his successors, the colony expanded to the west at the expense of the Indians until reaching the Appalachian Mountains. The Royal Proclamation of 1763 irritated colonists, but it was never effective because the territory was immense and British troops too few to enforce it, particularly when the empire had to concentrate its limited force in Boston. Hundreds of settlers pushed beyond the mountains to settle in the upper Ohio Valley during the late 1760s and early 1770s. The colony assumed a central role in the Revolution, and Philadelphia served as capital of the fledgling country before the move to Washington DC. Although the largest political plurality supported the Patriot cause, there were notable Loyalists also wrestled for control of the colony, including Joseph Galloway and William Allen. However, the majority of the citizens tried to stay out of the conflict.

One of the principal causes of political conflict within the colony was the disposition of land by the absentee Penn descendants. Although there was no royal governor in Pennsylvania, the Penn Family maintained the proprietorship and directly appointed the governor. The governor represented their interests, and the Penn Family continued to insist that their lands could not be taxed. Patriots, including the vociferous Benjamin Franklin, complained to the British government about this fundamental injustice, and Parliament's lack of response to the issue contributed to the growing sense of outrage in the colony.[2]

Delaware

In 1664, James the Duke of York conquered the area that became the state of Delaware from the Dutch. Over the next century, land disputes between Delaware and Maryland played out in London courts. Meanwhile, Pennsylvania's proprietor had leased the rights to the colony and integrated it into its government structure. The two colonies had the same governor but remained separate entities, and at the time of the Revolution, Delaware joined the Patriot insurgency and became its own state.

New Jersey

English colonists seized the area of New Jersey from the Dutch in 1664, and it became a proprietary colony controlled by James, Duke of York. He in turn gave the colony to two of his supporters, Sir George Carteret and Lord Berkeley of Stratton. Together they established a colony granting land to settlers and religious tolerance for all in exchange for quitrents paid annually. In 1702, East and West New Jersey became one united royal colony with a governor appointed by the king. Home to many Loyalists, including then Governor William Franklin, the illegitimate son of Benjamin Franklin, New Jersey eventually joined the Patriot insurgency and became a state.

New York

In 1664, the English conquered the Dutch colony of New Netherland, renaming it New York in honor of James II, Duke of York. German and English immigrants flocked to the colony and engaged in small-scale farming that produced foodstuffs for the British West Indies. The elites built up rich manors worked by tenant farmers along the Hudson River, and the colony prospered from its integration into the British economic system. At the time of the Revolution, two factions, each headed by an elite family, competed for power in the colony. The DeLancey faction aligned with the Empire, while the Livingston faction favored the Patriots. When compromise became impossible after Lexington and Concord, the Patriots seized power in New York.

Massachusetts Bay

The early Pilgrims, and later the Puritans, carved out successful settlements that comprised the heart of New England. From the start, the colony ran afoul of English trade regulation and religious policy. With the restoration of Charles II and the reign of James II, England clamped down on Massachusetts, vacating its charter and installing an unpopular governor canceled colonists' title to their lands to collect fees for new titles. In the Glorious Revolution of 1688-1689, the colonists overthrew the royal governor and resumed their former government. The Province of Massachusetts Bay received a new charter in 1691 with the combination of the Massachusetts Bay Colony, Plymouth Colony, the Province of Maine, Nantucket, and Martha's Vineyard. (It also included Nova Scotia until 1696.) Throughout the pre-Revolution period, the colonial assembly had contentious relations with most of the royal governors, each trying to assert its authority over the other. When the Stamp Act crisis occurred, the Patriots of Massachusetts became the leading voices for resisting the new taxes.

Politics within the colony were fractious, with prominent families, mostly in Boston, vying for control. After 1763, the royal governors bore the brunt of colonial dissatisfaction, and in the years leading up to the revolution, the friction between the king's governor and the peoples' legislature sparked into the flame that became the Patriot insurgency.

Rhode Island

Roger Williams first settled in what became Rhode Island in 1636 after he was banished from the Massachusetts Bay Colony for religious heresy. Williams and his associates (including Anne Hutchinson, William Coddington, John Clarke, Philip Sherman, and other dissidents) set out to establish a colony based on religious freedom in contrast to what they viewed as the oppressive orthodoxy of the Puritan colony to the north.

In 1644, the settlements of Providence, Portsmouth, and Newport joined together as the Colony of Rhode Island and Providence Plantations, governed by an elected council. The union of these three towns and Warwick received a Royal Charter in 1663.

Rhode Island became the first British colony to declare independence in May 1776 ahead of the Declaration of Independence. Virginia soon followed.

New Hampshire

Captain John Mason founded the colony that became New Hampshire in 1629. It was settled by English colonists most of whom were in the fishing business. The settlements eventually came under the jurisdiction of Massachusetts until 1679, when King Charles II split the two colonies. As revolutionary fervor began to spread, New Hampshire became the first colony to set up an independent government, although it delayed in actually declaring independence from Great Britain. The new state was the first to compose its own constitution.

During the 1760s, Benning Wentworth and his family ruled the colony. He maintained his influence primarily by dispensing land to key members of the Council, judiciary, and lower house, thus gaining their admiration and support. He protected the most important of industries—the lumber trade—which brought him popular support and the increasing skepticism of a distant Crown.[3]

Connecticut

The colony of Connecticut coalesced from several settlements under a royal charter granted in 1663. It became a conservative community that prospered from both farming and trade. Congregational churches contributed to its reputation for sober, quiet industry. The colony's leaders strongly supported independence in 1776, and the new state later became a stronghold for the Federalist Party.

NATIVE AMERICANS

> He [King George III] has excited domestic insurrections amongst us, and has endeavoured to bring on the inhabitants of our frontiers, the merciless Indian Savages whose known rule of warfare, is an undistinguished destruction of all ages, sexes, and conditions.
>
> —The Unanimous Declaration of the Thirteen United States of America, July 4, 1776.

> [The American Revolution was] the greatest blow that could have been dealt us, unless it was our total destruction...The Americans, a great deal more ambitious and numerous than the English, put us out of our lands...extending themselves like a plague of locusts in the territories of the Ohio River which we inhabit"[4]
>
> —Comments made in 1784 by an Indian emissary to the Spanish governor of St. Louis regarding the American revolution.

The consensus among historians is that the American Revolution was a disaster for Native Americans,[5] whether they supported the British monarchy or the Patriot cause. The revolution opened the floodgates to white settlement of native lands beyond the Appalachian Mountains. In general both patriot and loyalist officials took a condescending view of Native Americans, often referring to them as "savages." Nonetheless, both sides engaged with various Native tribes to secure their support, or at least their neutrality. Yet, it is necessary to appreciate the historical context of the relationship between whites and Native Americans to understand the alliance patterns that emerged during the revolution.

By 1776, Native Americans in what is now the eastern half of the United States had been interacting with Europeans for approximately two centuries. During this period, a significant amount of conflict between the two groups occurred, as Native Americans feared that European settlers sought to deprive them of lands they had occupied since time immemorial. However, the relationship between them was not entirely conflictual. In particular, mutually beneficial trade patterns emerged during this period, and in some instances, close proximity led to the establishment of close personal relationships between the two communities These economic and personal relationships in turn affected alliance patterns during the revolution.

By the 1760s, most native peoples had long become dependent on trade with European countries, so much so that they often regarded trade in existential terms where a cutoff was viewed as an act of war.[6] Metal goods such as iron or brass kettles, metal hoes, arrowheads, knives, guns, axes, and hatchets facilitated farming, cooking, and warmaking, and Indian tribes also obtained cloth from Europeans.[7] Some groups, such as Algonquian-speaking tribes in the northeast, had become so dependent on trade that they no longer produced stone tools and weapons, and in fact no longer had the capability to do so.[8]

Trade with Europeans had a similar effect on the Iroquois. Barbara Graymont, one of the leading historians of the Iroquois, noted that:

> The coming of the Europeans wrought profound changes in the lives and attitudes of the Iroquois. These Indians were in the Stone Age when the European settlers first met them and had developed skills that well adapted them to their way of life. With the acquisition of superior European metal implements, the Indians rapidly lost their old skills in fashioning stone and bone implements. The metal knives, aces, hoes, awls, needles, and kettles of the whites were fast becoming necessities. No longer did the Indian need to fell trees by the laborious method of girdling and burning. The nearly unbreakable brass and iron kettles meant an end to the formerly important occupation of pottery making. Muskets and rifles seemed to have become essential for hunting as well as warfare, and unfortunate indeed was the tribe that was lacking in these weapons when attacked by well-armed enemy Indians.

> As he lost his old skills, the Indian leaned more and more heavily on trade with the whites to secure his needs. In truth, not only did he now need these goods, but also he needed the white man with his skills to repair his guns and hoes and sharpen his axes. The frontier blacksmith and gunsmith performed an essential service not only for the Indians but for the white community in keeping the neighboring Indians loyal. It was no longer possible for the red men to go back to the old way of life, for the old way had been severely modified. The white man had now become a necessity for the Indian.[9]

In return for various goods, native groups provided Europeans with furs, including beaver, fox, otter, and lynx, which commanded high prices across the Atlantic as a result of overhunting in Europe.[10] As will be discussed in greater detail later, the superior ability of the British to furnish manufactured goods often factored heavily in alliance decisions made by various Native American groups during the revolution.

However, trade also played a role in highlighting power differentials between Native Americans and Europeans, which some have argued reflected centuries of differential development and structural conditions in Europe and North America in the centuries prior to the revolution. As noted by Taylor:[11]

> When the Europeans invaded, the Native Americans painfully discovered their profound technological and epidemiological disadvantages. They lacked the steel weapons and armor and the gunpowder that endowed the invaders with military advantage. Native peoples also could not match the wind or water mills that facilitated the processing of wood and grain. Lacking horses and oxen, native North Americans knew the wheel only in Mesoamerica as a toy. For maritime navigation, the natives possessed only large canoes and rafts incapable of crossing an open ocean in safety. Their lone domesticated animal was the dog, which provided far less protein and less motive power than the cattle and horses of the Europeans. Only the elites in parts of Mesoamerica possessed systems of writing

that facilitated long-distance communication and record-keeping. Consequently, in North America in 1492, only the Aztecs of Mexico constituted an imperial power capable of governing multiple cities and their peoples by command. In addition, no Native Americans possessed an ideology that impelled them far beyond their known world in search of new lands and peoples to conquer and to transform. Finally, compared to Europeans, the natives of America carried a more limited and less deadly array of pathogenic microbes.[12, 13]

By contrast, the Europeans of 1492 were the heirs to an older and more complex array of domesticated plants and animals developed about nine thousand years ago at the eastern end of the Mediterranean. The European mode of agriculture featured domesticated mammals – sheep, pigs, cattle, and horses – endowing their owners with more fertilizer, mobility, motive power, animal protein, and shared disease microbes. Building on a long head start and the power of domesticated animals, the Europeans had, over the centuries, developed expansionist ambitions, systems of written records and communication, the maritime and military technology that permitted global exploration and conquest, and (unwittingly) a deadly array of diseases to which they enjoyed partial immunities. Lacking those peculiar ambitions, technologies, diseases, and domesticants, the Indians did not expand across the Atlantic to discover and conquer Europe.

Profound civilizational differences and asymmetries conditioned and formed the backdrop to relationships between Euro-Americans and Native Americans. Therefore, by the time of the revolution, an overriding concern among native groups was to position themselves in alliances in such a way to maximize their independence and guarantee their survival in the face of growing settler encroachment.

Conversely, following the Seven Years War, a divergence emerged between colonial and imperial officials over the management of Native American affairs, and this difference played a prominent role in the outbreak of hostilities in the mid-1770s. The alliance between various Native American groups and the British established in the late 1750s proved vital in the British defeat of France in the Seven Years War,[14] which gained for the victors vast territories in North America. Such efforts at alliance formation and maintenance by the British required negotiation and bargaining with Native groups, which in turn obliged the former to take into consideration the interests of the latter, which typically revolved around land and access to trade.

On the other hand, for the colonists and, in particular, those on the frontier near Indian settlements, the main problem was not finding a mutually beneficial arrangement and alliance with native groups, but more fundamentally the Indian obstruction to further white expansion into the interior of North America.[15, 16] War was sometimes seen as a useful means for clearing Indian Territory for colonial settlement. For instance, following an Indian raid in May 1622 that killed nearly a third of the colonists in Virginia, Sir Francis Wyatt, the governor, stated that "Our first work is expulsion of the Salvages to gain the free range of the countrey for increase of Cattle, swine &c ... for it is infinitely better to have no heathen among us, who at best were but thornes in our sides, than to be at peace and league with them."[17]

For this group, therefore, the proper response towards Native groups did not involve negotiations and bargaining but rather their elimination or removal to reduce the threat from raids and to open up territory for settlement. This tension between the interests of imperial and colonial officials was captured by Graymont, who noted that:

> Those who lived closest to the Indian territories and who had no more far-reaching responsibilities than their own self-interest or the interests of their local communities had frequently looked upon the Indian tribes from an entirely different perspective from that of the British administration. Many of the colonial leaders and border settlers considered the Indian to be an inconvenient impediment in the way of further white advancement – a creature to be appeased only when absolutely necessary, to be exploited always, and to be divested of his land whenever feasible. His Majesty's government and the crown officials charged with Indian administration were thus often strongly at odds

with the local colonial leadership and the frontiersmen in the matter of Indian affairs.[18]

To many colonists, it seemed that the Crown had sided with Indian allies over its colonial subjects when it issued the Royal Proclamation of 1763, which prohibited white settlement west of the Appalachian Mountains.[19]

With the removal of the French from North America following the Seven Years War, and because accommodating Native interests was so harmful to Imperial-Colonial relations, why did imperial officials believe accommodation was practicable? In more blunt terms, why didn't imperial officials agree with the colonial assessment that the optimal Indian policy was one of, in modern parlance, ethnic cleansing? Such a possibility was recognized by Taylor, who noted that:

> British success [in 1763] threatened the Indian peoples of the interior, for they depended upon playing off rival empires to maintain their own autonomy. Deprived of a French counterweight, the British Empire could sweep settlements deep into the continent, pushing the Indians aside and transforming their lands into farms and towns. The victory of the British introduced a dangerous new stage in the colonization of North America, when their large colonial population on the Atlantic seaboard would break through the Appalachian Mountains into the great heartland of the continent, the vast Mississippi watershed.[20]

Some imperial officials may indeed have harbored such a preference. Schmidt noted that "Lord Jeffrey Amherst, now in command of all British forces in North America, moved quickly to demonstrate to western Indians that at best he viewed them as subjects to be commanded and at worst as nuisances worthy of extermination."[21] Additionally, some native groups harbored fears that ethnic cleansing was indeed the preferred British policy. In correspondence with John Stuart, the Crown's Superintendent of Indian Affairs for the southern district of North America, the Earl of Shelburne, the Secretary of State for the Southern Department, noted in December 1766 that it was "much more preferable . . . to gain the Affection of those tribes by mild Treatment, than to set them at Variance with each other, which ultimately must confirm them in the impression which they have already entertained, that we wish the Extirpation of them all."[22]

The simple answer is that by the end of the Seven Years War financially the British were not in a position to undertake a major military campaign in North America to clear territory of Indian inhabitants to pave the way for colonial settlement. The war increased British debt from £73 million to £137 million, while interest payments on the debt amounted to 60 percent of budgetary expenditures in the 1760s.[23, 24] Additionally, Pontiac's rebellion, in which the British were forced out from a number of forts in the *pays d'en haut* (i.e., "upper country" in the Great Lakes region), led to the deaths of 450 British soldiers and the dispersal of four thousand colonists from the frontier. Such actions revealed to the British that Native groups working on concert had the ability to impose prohibitive costs on the empire.[25, 26]

These constraints factored heavily in the belief of representatives of His Majesty's government managing Indian affairs that imperial interests were best served by taking as given the presence of Indian communities between the Appalachians and the Mississippi river, and to which at least some measure of legitimacy had to be afforded to the power, interests and territorial claims of various native groups.[27, 28] With the option of population transfer simply out of the question, the British adopted a policy of *mission civilisatrice* and enlightened paternalism toward native tribes, in which the British would steer Indians from barbarism to civilization. Schmidt noted that:

> [T]he British government saw Indians in the west as inferior people, but people nonetheless who were capable, with the proper instruction, of ascending to "civilization." At the same time, British imperial administrators, politicians, and officers also wholeheartedly believed that the lands of the Indians of the west were solely the property of the British Empire by virtue of their victory in the Seven Years' War, so they therefore felt entirely justified in taking possession of the Ohio Valley and in forcing their version of civilization on the area's inhabitants. In the British imperial view, the Indians of the Ohio Valley possessed neither civilization nor sovereignty, and the British, by virtue

of their possession of both, would bring Ohio Valley peoples to civilization.[29, 30]

Native groups did not believe they occupied their territory only through British sufferance. Yet, as noted by Schmidt, British policy encapsulated a fundamental contradiction. Great Britain assured western tribes of the validity of the latter's claim to land over which they simultaneously claimed British sovereignty. This contradiction encouraged colonists to disregard Indian territorial claims.[31] Indeed, many colonists believed that the Seven Years' War was fought to obtain the Ohio valley for their settlement.[32] It is perhaps not surprising, therefore, that policies preferential to native interests, such as the Royal Proclamation of 1763, were seen by colonists as infringements on autonomy and self-government and in fact as part of a larger plot hatched by imperial officials in London and North America intended to destroy freedom in the colonies.[33, 34] Yet, despite the contradiction, most Native groups concluded that tribal interests concerning autonomy and territorial claims were better served by allying with the British rather than the Americans. From this perspective, the tensions leading up to the revolution, particularly along the frontier, can perhaps be viewed as a tripartite struggle, with a majority of Indian groups siding with the British against patriot forces.

In the early sixteenth century, a prophet known as Deganawidah and his chief disciple Hiawatha convinced the Five Nations of the Iroquois, specifically the Mohawks, Oneidas, Onandagas, Cayugas, and Senecas, to form the Great League of Peace and Power, which helped put an end to terrible wars the different tribes waged on each other.[35] The Great League thereby provided an alternative to the "mourning wars" the various tribes often engaged in, by providing a setting whereby condolence ceremonies and ritualistic gift-giving replaced the usual practice of revenge killings and the taking of prisoners to replace those killed.[36, 37]

By the seventeenth century. the Great League evolved into the Iroquois Confederacy in response to the arrival of Europeans. In addition to essentially amounting to a mutual nonaggression pact between the five constituent tribes, it served as a forum to harmonize foreign policies while permitting autonomy over local affairs.[38] In the early eighteenth century, the Tuscarora became the sixth nation within the Confederacy after they were driven out of North Carolina, and up until the Seven Years War, most of the constituent nations of the Confederacy had Anglophile, Francophile, and neutralist wings.[39]

In the 1670s, the English governor of New York, Sir Edmund Andros, established an alliance with the Iroquois known as the Covenant Chain, through which the Iroquois were able to influence British diplomatic and, at times, military policy toward other northeastern native groups,[40] which in turn invited the Iroquois to dominate neighboring tribes.[41] Levinson noted that by the time of the revolution, the Iroquois were perhaps the most formidable military organization in New York State, with a capacity to field two thousand warriors (out of a total population of ten thousand).[42] However, although both Loyalists and Patriots courted the group, the Iroquois initially took a neutral stance in the conflict between Crown and colonists, which was reflected in comments made in August 1775 by the Mohawk chief Tigoransera, otherwise known as Little Abraham: "This, then, is the determination of the Six Nations…Not to take part, but, as it is a family affair, to sit still and see you fight it out … for we bear as much affection for the King of England's subjects, upon the other side of the water, as we do for you, born upon this island."[43]

However, one year later, by the summer of 1776, the revolution had split the Iroquois Confederacy, with most of the Mohawk, Seneca, and Cayuga nations supporting the English, the Oneida and Tuscarora supporting the Americans, and the Onondaga holding on to neutrality.[44] Concerns over land and colonial encroachment weighed heavily on alliance decisions. The most pro-British faction were the Mohawks, and an important figure mediating the Mohawk relationship with the British was the Mohawk chief Thayendanegea, or Joseph Brant, who was the younger brother of Molly Brant, the wife of Sir William Johnson, the late Superintendent of Indian Affairs for the Northern District. Yet Brant's affinity for the British was likely established earlier as a result of his education at the Anglican Mohawk Mission, and he was a devout Anglican.

While Brant's upbringing and attachment to English leaders played an important role in determining his allegiance during the revolutionary war, his pro-British stance was also based on the belief that Mohawk territorial claims could best be secured through an alliance with the British. Brant believed that the Americans desired Mohawk lands, that they "began this Rebellion to be sole Masters of the Continent," and that they would not honor pledges made to the Mohawks if they defeated the British.[45] Instead, Brant sought British assurances to respect Mohawk territorial claims, including from King George on a trip to England, and in 1779, he obtained a captain's commission from the British and promises that all Mohawk lands would be returned to them if the British defeated the Americans.[46]

Concerns of American dispossession of native lands also motivated the attitudes of other Iroquois groups. For instance, in 1779, the Seneca chief Sayengeraghta, known as Old Smoke, in discussing the defense of Fort Niagara, noted that:[47]

> It is also your Business Brothers to exert yourselves in the Defense of this Road by which the King, our Father[48, 49] so fully supplied our Wants. If this is once stopt we must be a miserable People, and be left exposed to the Resentment of the Rebels, who, notwithstanding their fair Speeches, wish for nothing more, than to extirpate us from the Earth, that they may possess our Lands, the Desire of attaining which we are convinced is the Cause of the War between the King and his disobedient Children.

Most of the Oneida and Tuscarora supported the Patriots, although initially the former were reluctant to take sides. In a March 1775 speech to the governor of Connecticut, the Oneidas stated that "We are unwilling to join on either side of such a contest, for we love you both – old England and new. Should the great King of England apply to us for our aid – we should deny him – and should the Colonies apply – we shall refuse."[50]

The latter eventually turned out not to be the case, as the Oneidas did eventually support the patriot cause. This decision is often attributed to the influence[51] of Samuel Kirkland, a Presbyterian minister with pro-patriot sympathies who lived among the Oneidas from 1766 until his death in 1808. Indeed, writing in 1775 to patriot leaders in Albany, Kirkland noted that his description of the conflict in a manner favorable to the patriot cause had "done more real service to the cause of the Country, or the cause of truth and justice, than five hundred pounds in presents would have effected."[52]

While noting the impact of Kirkland, Levinson also attributed the Oneida-Colonist alliance to other factors, including political, economic, and social ties formed by the Oneida over many years with nearby settlers who supported the revolution.[53] Beginning in the early eighteenth century, missionaries played an important role in teaching English to the Oneida and converting many tribal members, including tribal chief Peter Agworondougwas, known as "Good Peter," to Christianity, both of which helped foster ties with local settlers.[54] Additionally, by the outbreak of the revolution, the Oneida had become economically integrated with local settlers, trading furs and meat in return for tools, farm equipment, cooking materials, guns, coffee, tea, medicine, bread, and flour. Hence, British policies detrimental to local settlers, such as those on taxation, also impacted the Oneida and thus shaped their attitudes toward the Crown.[55, 56]

The Oneida aided the colonists in a number of capacities during the revolution. Notably, in addition to serving as guides and interpreters, and as diplomatic representatives to pro-British factions within the Confederacy, the Oneida also served as informers and spies.[57] Specifically, in the summer of 1775, although officially neutral, they reached an agreement with the Committee of Safety of Tyron County to "communicate to us the Committee of Safety all the Remarkable News and Intelligence they can get in Regard to these present Troubles and desired the same of us reciprocally."[58, 59] Matters came to a head within the Iroquois Confederacy in 1777 with the Battle of Oriskany, which featured a force of seven hundred British and loyalist troops supported by eight hundred fighters from the Seneca, Cayuga, Onondaga, and Mohawk nations (along with several non-Iroquois tribes), arrayed against a patriot force featuring Oneida and Tuscarora warriors. The sanguinary encounter resulted in a British victory, and the clash among the Iroquois warriors led to retaliatory raids against each other's villages and crops.[60, 61]

In the seventeeth century, Puritan missionaries established multiple "praying towns" in Massachusetts that attracted various native peoples interested in converting to Christianity and adopting Puritan ways. One such town, Stockbridge, was established in 1736 in western Massachusetts. It was initially populated by ninety Native Americans, primarily Mohicans but also eventually Wappingers.[62] They were taught English and attended church services conducted in Mahican, and town residents became members of the Congregationalist Church, the same denomination as that of many revolutionary leaders throughout New England.[63] Such religious ties, as well as social bonds formed with patriot sympathizers among neighboring settlers likely played a strong role in determining Stockbridge allegiances during the revolution. At the Treaty of Albany in August/September 1775, Captain Solomon Uhhaunauwaunmut told the Americans, "Wherever you go we will be by your Side Our Bones shall lay with yours. We are determined never to be at peace with the Red Coats while they are at Variance with you… if we are conquered our Lands go with yours, but if we are victorious we hope you will help us to recover our just Rights."[64, 65]

Stockbridge Indians served with colonial forces at different times during the revolution. Seventeen were with Washington's forces in Cambridge during the spring of 1775, and they also served with Horatio Gates' army at White Plains in the summer of 1778.[66] Notably, the Stockbridge suffered heavy losses in area that is now Van Cortlandt Park in the northern Bronx. The Wappinger chief Daniel Nimham fought in this battle, with one Hessian soldier noting in his diary that "No Indians, especially, received quarter, including their chief called Nimham and his son, save for a few."[67]

For Calloway, the participation of the Stockbridge on the side of the Patriots provided the British with the excuse they needed to mobilize Indians in support of the Crown. In September 1775, General Thomas Gage, who had replaced Amherst as commander of all British forces in North America, writing from Boston to Stuart, the Crown's Superintendent of Indian Affairs for the southern district of North America, noted that "the Rebells have themselves open'd the Door; they have brought down all the Savages they could against us here, who with their Rifle men are continually firing on our advanced Sentries."[68]

Interestingly, there is some uncertainty among historians as to why the Stockbridge supported the patriot cause,[69] particularly because Stockbridge Indians were forced to sell most of their land to whites to pay off debts and to support themselves. In 1763, Native Americans in Stockbridge owned three quarters of the town's land, most of which was held communally, while by 1774 this figure was reduced to 6 percent, and a similar pattern was observed in other praying towns.[70] Schmidt suggested that the Stockbridge might have acted strategically, in that fighting against patriot sympathizers among their neighbors would only have hastened their total dispossession, and so the interests of the group were better served by enhancing links with their neighbors, rather than putting their faith in distant leaders in London.[71]

The Cherokee separated from Iroquoian groups about 3,500 years ago,[72] and by the mid-eighteenth century, they numbered about twelve thousand inhabitants in approximately forty villages in the southern Appalachian Mountains.[73] They were organized along two major divisions, specifically the Overhill towns along the Little Tennessee, lower Hiwassee, and lower Tellico rivers in Tennessee, and the Lower towns along the upper Savannah River in South Carolina.[74]

Calloway noted that Cherokee population totals were higher earlier in the century (at twenty-two thousand), with smallpox epidemics and participation in European wars in North America contributing to depopulation.[75] Another factor was the Anglo-Cherokee war of 1759-1761, which featured Cherokee raids along the South Carolina frontier and the capture of Fort Loudon, as well as the destruction of Cherokee towns and crops by British and colonial forces in 1761. The Cherokee were forced to sue for peace, ceding large amounts of land to the British. Additional Cherokee land was signed away by the Iroquois in the Treaty of Fort Stanwix prior to the revolution, and further dispossession occurred as a result of trade, which saw Cherokee Indians run up large debts that could only be discharged through the sale of land.[76]

By the time of the revolution, therefore, Cherokee grievances had built up over the loss of land to white settlers. Such grievances were particularly acute among young male Cherokees who felt that older leaders, such as Attakullakulla, or Little Carpenter, had bargained away their patrimony to maintain trade links and to stem the tide of settler encroachment by establishing new borders to contain backcountry settlement.[77] The timing of the revolution was therefore ideal for young leaders such as Tsi'yugûnsi'ny, or Dragging Canoe, Attakullakulla's son, as it offered an opportunity to seek historical redress through the reclaiming of land recently lost to Euro-American settlement. In a meeting with Henry Stuart, the brother of the Superintendent of Indian Affairs, Dragging Canoe complained that the Cherokee "were almost surrounded by the White People, that they had but a small spot of ground left for them to stand upon and that it seemed to be the Intention of the White People to destroy them from being a people."[78]

War became unavoidable in April 1776 following the arrival of a delegation of Shawnees, Delawares, Mohawks, Nanticokes, and Ottawas to the Cherokee town of Chota. Painted in black, the delegates urged a unified resistance to the Americans and presented a nine-foot long wampum war belt to Dragging Canoe, who accepted it. In regards to Atakullakulla and other chiefs, such as Oconostota, Stuart noted that, "Instead of opposing the rashness of the young people with spirit, [they] sat down dejected and silent."[79]

The Cherokee War of 1776 proved calamitous for the tribe. Dragging Canoe and his followers attacked into South Carolina, in July 1776 killing sixty settlers, instilling mass panic on the frontier. Notably, the attacks did not distinguish between Loyalists and Patriots, thereby indicating that the objective of the campaign was not to preserve British rule in North America but to roll back white settlement of the frontier.[80]

The response was deadly. With the support of General Charles Lee, the Continental Army's commander of southern forces, militias from the Carolinas and Virginia laid waste to Cherokee territory in the late summer and early fall of 1776, and the Cherokee were largely defeated by November. Peace talks were conducted in the spring, and in May, the Lower Cherokees signed away most of their claims to South Carolina in the Treaty of Dewitt's Corner. In July, the Overhill Cherokees were forced to renounce their territory east of the Blue Ridge Mountains and combined these cessations amounted to over five million acres of territory.[81]

Dragging Canoe and his followers, though, remained unreconciled and retreated to territory near present-day Chattanooga. As a form of protest, they abdicated their Cherokee identity and began to refer to themselves as "Chickamaugas,"[82] after the name of the creek where they encamped, and derisively referred to those who did not join them as "Virginians."[83]

The Chickamaugas continued to attack frontier settlements into 1777, and they assisted British forces in the capture of Savannah and Augusta in 1778 and 1779.[84] In 1780, Thomas Jefferson, then governor of Virginia, ordered a devastating campaign against Cherokee territory, which in the words of one Cherokee leader, saw the Virginians "dyed their hands in the Blood of many of our Women and Children, burnt 17 towns, destroyed all our provisions by which we & our families were almost destroyed by famine this Spring."[85] Fearful of being completely dispossessed, the other Cherokees attempted to divert patriot attacks toward the Chickamaugas. Toward the end of the war, Dragging Canoe attempted to establish a broad federation of Indian tribes consisting of Cherokee, Creek, Chickasaw, Seminoles, and others to resist the Americans, but such plans were pre-empted by the Treaty of Paris, which led to the termination of British support for Indian resistance efforts.[86] Nonetheless, Dragging Canoe and the Chickamaugas continued the struggle, now against the United States, into the 1790s.

Land and access to trade figured prominently in Creek calculations before and during the revolution. The Creeks were organized along two main divisions, with the Lower Creeks situated in southwestern Georgia near the lower Chattahoochee, Flint, and Ocmulgee rivers, and the Upper Creeks in Alabama along the Tallapoosa, Coosa, and Alabama rivers. Prior to the revolution, the Lower Creeks ceded over two million acres of land to Georgia in 1763 during the Congress of Augusta, a transfer initially opposed by the Upper Creeks.[87] Lacking support from either the French or Cherokees, the Upper Creeks backed the cession the following year, and in 1765, both factions agreed to a land cession to the British in West Florida. To the consternation of many Creeks, a further cession of over two million acres occurred in 1773, as the Creeks were forced to cede land to repay exorbitant trade debts.[88, 89]

Both the British and Americans courted the Creeks during the revolution, with both sides seeking to keep the tribe neutral and outside the orbit of the other side. The Americans were able to make some inroads with the Lower Creeks shortly before the revolution, as the latter were swayed by the decision of the South Carolina Council of Safety to provide trade and gifts to the faction.[90] The Patriots also won over the Upper Creek leader Handsome Fellow, through the provision of presents[91, 92] and promises to exonerate a Creek warrior accused of murder.[93]

Yet other factors mitigated against American efforts to keep the Creeks neutral. First and foremost was the Creek concern that a patriot victory would open the floodgates to further settler encroachment and dispossession of Creek lands.[94] A second factor was the close kinship, economic and military relationships between Creeks and Cherokees, and the decimation of the Cherokees in 1776 led to a strong desire for revenge among the Creeks. The Americans, though, recognized the deterrent effect of the defeat of the Cherokee, with General Arthur Lee warning the Creeks they would face a level of destruction on par with that of the Cherokee if they abandoned neutrality.[95] David Taitt, John Stuart's deputy, acknowledged the importance of this threat in noting that "The fate of the Cherokees has struck the people of this nation with such a panic that, although they have a great aversion to the rebels yet they are afraid to go against them until they hear of his Majesty's troops being at Charlestown or Savannah."[96]

Throughout the war, both neutralist and pro-American factions competed with pro-British supporters for control of Creek allegiance. However, increasing settler encroachment and attacks on Creeks, and the British ability to provide superior goods and trade undermined American efforts to sway the tribe.[97] Although the war had disrupted British trade at various times, by 1778, the renewed supply of British goods gave the pro-British faction led by Alexander McGillivray the upper hand, and by the summer Creeks (and Seminoles) were defending British outposts and participating in assaults against American forces.[98] Creeks also fought the American siege of Savannah during 1781 and 1782, but by the end of the war, the pro-British faction lost sway within the group as British trade and goods disappeared. At this point, the pro-American faction was ascendant in Creek politics, and in November 1783, the tribe signed a peace treaty with Georgia, which saw it cede eight hundred square miles of territory along with authority to regulate trade.[99] This agreement, though, was not recognized by the pro-British faction led by McGillivray.

The Seminoles were not a significant player during the revolution, which can be accounted for in part by noting the group's incomplete ethnogenesis by the outbreak of hostilities. That is, the tribe that went by the name "Seminole" did not exist prior to 1750 and was not an entity independent of the Creeks until after the revolution.[100] The Seminoles were originally Creek migrants from central Georgia who moved south into territory that would become Florida after the British acquired the land in 1763.[101] Seminole leaders often took their cues from the Creeks. For instance, in 1776, Governor Patrick Tonyn of East Florida sought to recruit the Seminoles for a campaign against Georgia, yet the group hesitated to commit themselves until they received instructions from the Lower Creeks.[102]

Seminoles in the western part of the state near Tallahassee favored the Spanish, who held Louisiana, while those further east near Alachua were strongly pro-British, and an important figure for this latter group in maintaining the British relationship was Ahaya of Cuscowilla, or Cowkeeper. By the mid-1770s, Cowkeeper had long maintained a pro-British orientation, as in 1764, he indicated to Stuart that he "would always hold the white people fast, even if their Nation should behave otherwise."[103] As in the case of the Creeks, trade appears to have been an important factor in Seminole sentiments, as by the end of the revolution, the Seminoles were integrated into an Atlantic trading system that furnished them a variety of goods, including Welsh plain cloth, Yorkshire broadcloth, Irish linen, saddles, shoes, smooth-bore muskets, iron pots, kettles, pans, axes, and carpenter tools, among other goods.[104] The Alachua Seminoles were firmly pro-British throughout the war.[105]

The Shawnees were one of the more prominent tribes in the Ohio Valley, and they consisted of five main divisions, specifically the Chillicothe and Thawekila division, which typically handled political matters affecting the whole tribe; the Piquas, who managed religious affairs and rituals; the Maquachakes, whose portfolio included health and medicine; and the Kispokis, who handled war preparations and supplied war chiefs.[106] In the lead up to the revolution, the tribe had a fairly troublesome relationship with colonists seeking to settle on Shawnee lands along the frontier. In the 1768 Treaty of Fort Stanwix, the Iroquois ceded Shawnee territory south of the Ohio River, in what is today the state of Kentucky, to the British. This ploy by the Six Nations represented an effort to steer settler land hunger away from Iroquois territory further north.[107] Although the Shawnee had long acknowledged the Iroquois as "elder brothers," such recognition of Six Nation status did not extend to granting them the right to negotiate away Shawnee territory.[108] Hence, it was not until six years later when the Shawnee reconciled themselves to the loss of the territory south of the Ohio River following defeat in Lord Dunmore's[109, 110] war to a Virginia militia force led by Colonel Andrew Lewis.

Territorial pressures were nothing new to the Shawnees. Defeat at the hands of the Iroquois in the latter half of the seventeenth century led to the dispersal of the tribe to Pennsylvania and to French settlements in Starved Rock and Fort St. Louis in Illinois country and further south along the Savannah River.[111, 112] This latter group participated in the Yamassee War of 1715, a conflict that threatened the existence of the colony of South Carolina and that was waged by native groups to halt encroachments by settlers from Virginia and the Carolinas. The aftermath of this conflict saw the southern group of Shawnees move once again, some further south and west into Creek territory, while others went back north into Pennsylvania.[113] By the mid-eighteenth century, most Shawnees were back in the Ohio Valley, yet the aftermath of the Seven Years' War brought a flood of colonists and traders into the Ohio Valley, along with an increased British military presence.[114, 115]

The British military presence was minimal and was withdrawn in 1772. Moreover, by this time, most of the Shawnee had been relocated elsewhere,[116, 117] and thus, in the lead up to the revolution, encroachment by frontier settlers was of paramount concern. Hence, in July 1775, one Shawnee chief informed a group of Virginians that, "We are often inclined to believe there is no resting place for us and that your Intentions were to deprive us entirely of our whole Country."[118]

None of the major Shawnee divisions seriously entertained the notion of supporting the Americans, although the Maquachakes, under the leadership of Cornstalk, had advocated a policy of neutrality.[119] Cornstalk had also opposed participation in Lord Dunmore's war, yet once hostilities began, he helped organize Shawnee defenses,[120] as some Virginian troops recalled hearing him urge his troops to stand their ground during battle.[121] The neutralist stance appears to have been based on the recognition of the tribe's dependency on Anglo-American trade and resignation to the loss of Kentucky.[122] However, the Shawnees were troubled when Kentucky became a county in Virginia in the fall of 1776, and concerns over settler encroachment were evident in the following comments made by Cornstalk in November 1776 to the Second Continental Congress:

> Our Lands are covered by the white people, & we are jealous that you still intend to make larger strides. We never sold you our Lands which you now possess on the Ohio between the Great Kanawha & the Cherokee, & which you are settling without ever asking our leave, or obtaining our consent…Now I stretch my Arm to you my wise Brethren of the United States met in Council in Philadelphia. I open my hand & pour into your heart the cause of our discontent in hopes that you will …send us a favorable Answer, that we may be convinced of the sincerity of your profession.[123]

Cornstalk advocated neutrality despite these concerns, and in fact in the summer of 1775, he had informed the Americans that many Piquas opposed neutrality.[124] However, all possibilities for neutrality evaporated with the murder of Cornstalk in 1777 at Fort Randolph by militiamen stationed at the fort. The Maquachake chief had traveled to the fort to inform the Americans of the increasing radicalization of members of his tribe who were eager for war against Virginians.[125] Yet once in the fort, he was held captive and then murdered by those seeking revenge for the killing of members of a patrol by Indians allied with the British.

The murder of Cornstalk simply sped up the development of trends that would have occurred otherwise. Indian hatred was widespread along the frontier, as was the desire for Indian Territory. George Morgan, an Indian agent for the Americans with responsibility for tribes in the Ohio Valley, in discussing the attitudes of settlers near Fort Pitt, noted an "ardent desire for an Indian War, on account of the fine Lands those poor people possess."[126] For militant Shawnees, the revolution offered an opportunity to roll back white settlement, and in 1777, the Shawnee Chief Blackfish invaded Kentucky territory and captured the frontiersman Daniel Boone and later attacked Fort Randolph. Meanwhile, American incursions into Shawnee territory and the burning of crops and villages were a common feature in the late 1770s.[127] Kentucky settlers also lived under the fear (and reality) of Indian attacks, while word of atrocities committed by "the White Savages Virginians" spread throughout Indian Territory.[128]

Shawnees constituted a large component of the 1,200 troops led by British officer Henry Bird that attacked various Kentucky settlements in June of 1780. The settlements fell to Bird's forces, who took 350 captives, two hundred of which were given to Indian allies while the rest were marched back to Detroit, which was then under British control. However, along the way a number of captives who were unlikely to survive the journey, including women and children, were tomahawked, perhaps out of a sense of mercy by those natives who sanctioned and committed the killings.[129] To avenge this atrocity, in August, the Virginian George Rogers Clark led a force of nearly one thousand men against the Shawnee town of Chillicothe, which was abandoned several hours prior to Clark's arrival. Clark's force proceeded to burn Chillicothe and then marched to northeast to Piqua, where they defeated a force of three hundred Shawnees, Delaware, Mingo. and Wyandotte Indians.[130] Clark's force had to return to Kentucky due to a lack of supplies, but he invaded Shawnee territory again in 1782,[131, 132] and Calloway noted that by this time the conflict had become a total war for the tribe, as children as well as old men and women had taken up arms.[133]

The American Revolution and its aftermath turned out to be a disaster for the Shawnees, as between 1775 and 1790, eighty thousand settlers moved into Shawnee territory.[134] This development highlights a more general observation about the revolution in territory beyond the Appalachians. Whereas abstract principles related to liberty and representative government motivated Patriots in Boston, Philadelphia and elsewhere back east, the conflict in the Ohio Valley was largely a vicious racially tinged struggle over territory that preceded and became embedded within the revolution, yet remained distinct in its motivating impetus from the broader enveloping conflict with Britain (and in fact continued following the Treaty of Paris[135, 136]).[137, 138] As noted by Schmidt:

> As the American Revolution began, the attention of both colonists and Indians in the Ohio country was focused much less on issues of taxation, representation, and Parliamentary versus local control than on the results and repercussions of the recently concluded Dunmore's War. Despite their growing disagreement regarding the power of the legislature versus that of the Royal governor, Virginians of all ranks concurred with Dunmore that the purpose of the conflict with the Ohio Valley Indian groups had been to establish once and for all, "an idea of the power of the White People upon the minds of the Indians."[139]

Finally, one group of Shawnees supported a neutrality that leaned toward the Americans, specifically a collection of Shawnees (and Delaware Indians) who inhabited villages around Coshocton. Led by the Delaware Indians White Eyes and Killbuck, this was the only group of Indians in the Ohio Valley that supported the Americans, and their alliance decision appears to have been based on several main factors. First, neither leader believed that a conflict with American settlers could be won, even with British assistance.[140] Second, White Eyes sought Congressional recognition of Delaware territory north of the Ohio River. In particular, he sought to secure Coshoctan's independence under a moderate form of American suzerainty. He envisioned his people would "[live] as White people do under their Laws and Protection," in exchange for technological and economic benefits of Euro-American civilization.[141]

Neither White Eyes nor Killbuck agreed to provide active military support to the Americans, although they were willing to guide American forces toward their enemy and to negotiate on their behalf with other Indians groups.[142] However, the policy of soft alignment with the Americans came under severe strain with the killing of White Eyes by American militiamen and various other killings of Indians soured relations and led some Coshocton Indians to join their more militant brethren. By July 1779, however, Killbuck and the Coshocton council agreed to become allies of the United States, and in August, eight Coshocton warriors participated in an attack on the Seneca nation.[143] Not surprisingly, later that fall the Coshocton Indians were disowned by various chiefs and warriors from the Shawnee, Delaware, Mingo, and Huron tribes.

The Chickasaw were a small tribe of around 2,300 people (including 450 warriors) located in northern Mississippi and western Tennessee. During the late seventeenth and early eighteenth centuries, the tribe had allied with the British and against the French, and the foundation of the alliance rested on trade, as the British provided large quantities of guns, ammunition, and other metal weapons in return for deerskins and Indian slaves obtained through raids on neighboring villages.[144] Weapons obtained from the British in turn enabled the small tribe to maintain a degree of independence from larger tribes.

As in the case of other tribes, the main concern of the Chickasaws leading up to the revolution was the increasing presence of white settlers, squatters, and traders on their territory. The increasing presence of the first settlers—which the natives viewed as the same as squatters—put pressure on the carrying capacity of the local environment, as now local supplies of plants, land, and animals had to support a growing population.[145] Chickasaw leaders also complained of the activities of colonial traders, whose growing influence among the tribe and sales of rum were seen by Chickasaw leaders as corrupting the tribe and undermining their influence.[146]

The Chickasaws proved to be hesitant allies of the British during the revolution. They made it clear to John Stuart, the British Superintendent of Indian Affairs, in meetings held in May and June of 1777 in Mobile, that their support would be forthcoming only if the British could guarantee the maintenance of trade free from deception and fraud.[147] Chickasaw hesitancy also stemmed from the fact that some members of the tribe had established close relationships with colonists living in the frontier.

In describing a conversation with Chickasaw chief Paya Mataha, Stuart noted that:

> In the course of my conversations with the chief, I found that it was with the utmost difficulty he could place in the light of enemies those men whom from his earliest infancy he had been taught to consider as his dearest friends, whom he had assisted and defended upon many occasions at the risk of his life. I had also the greatest difficulty to make him comprehend that they had forfeited their right to the protection of the Great King and the British nation by their apostasy and rebellion; and he at last observed that although these might be considerations of sufficient weight to engage us to make war upon them, yet he could not bring himself to imbrue his hands in the blood of white people without the greatest reluctance, and that he shuddered at the apprehensions of committing some fatal blunder by killing the King's friends instead of his enemies.[148]

At various times, American forces had threatened to invade Chickasaw territory, and the tribe had often received warnings from native groups further north that the Americans intended to destroy the tribe and repossess their land.[149] The tribe was unmoved, and following a May 1779 message from Virginia offering the choice of friendship or destruction, tribal chiefs responded as follows:

> We desire no other friendship of you but only desire you will inform us when you are Comeing and we will save you the trouble of Coming quite here for we will meet you half Way, for we have heard so much of it that it makes our heads Ach, Take care that we don't serve you as we have served the French before with all their Indians, send you back without your heads. We are a Nation that fears or Values no Nation as long as our Great Father King George stands by us for you may depend as long as life lasts with us we will hold him fast by the Hand.[150]

Needless to say, the tribe did not choose friendship with the Americans, a key factor being the tribe's enmity toward the French (who they had fought against, along with tribes allied to the French, several times in the eighteenth century). Chickasaws aided the British by patrolling the Mississippi river with the Choctaws, and they also helped defend Pensacola from the Spanish. However, the rationale for the alliance evaporated with the drying up of British trade following the surrender at Yorktown.[151]

The *pays d'en haut*, or the upper country, located around the Great Lakes region, comprised one of the two main regions of the colony of New France, the other being the St. Lawrence Valley.[152] Attacks by the Iroquois against various native peoples in the eastern Great Lakes region in the 1640s and 1650s led to an outflow of refugees westward, where they settled just south of Lake Superior, west of Lake Michigan, and along the upper reaches of the Illinois river.[153] The refugees consisted of many different native groups that collectively became known as the *pays d'en haut*, including the Fox, Huron-Petuns, Mascouten, Kickapoo, Ojibwa, Sauk, Potawatomi, Miami, Wyandot, Ottawa, Winnebago, Menominee, and Illinois, many of which were Algonquin speakers.

The French went westward from the St. Lawrence Valley following the destruction of the Huron by the Iroquois in the 1640s, which disrupted the fur trade. In the 1670s and 1680s, French traders and priests crossed the Great Lakes in canoes, and the French established important outposts in Cahokia, Kaskaskia, Vincennes, Detroit, and Michilimackinac (Quebec served as the colonial capital of New France).[154] By 1750, approximately two thousand French traders and settlers lived in the upper country among eighty thousand Indians.[155]

In addition to needing the cooperation of the *pays d'en haut* to carry out the fur trade, the French also needed them as allies against the Iroquois, whom the French believed were under British sponsorship.[156] Because neither the French nor the Iriquois could achieve their objectives through force alone, the two groups developed what the historian Richard White has called "the middle ground," characterized by a set of practices and discourses based on creative inventions and misunderstandings, that facilitated cooperation.[157] One such creative misunderstanding was the different conceptions of the "father" role that the French assumed in their relationship with the *pays d'en haut*.

The French realized that the only way they could mobilize the *pays d'en haut* to serve as allies against the Iroquois was to assume a mediator role when conflicts arose among the groups constituting the *pays d'en haut*. As one French Jesuit noted, "It is absolutely necessary to keep all these tribes . . . in peace and union against the common enemy – that is, the Iroquois."[158] Agreements among native groups were solidified and legitimated through the provision of gifts, which the French could provide. The various groups of the *pays d'en haut* simultaneously viewed the French as allies, protectors, mediators, and suppliers and referred to the French governor of Canada as "Onontio," an Iroquois word signifying "great mountain."[159] The *pays d'en haut* viewed the French as the leaders of the alliance, and within this context, the French indicated that the King of France was the "father" to the native "children."[160] Such terminology implied the French viewed the relationship with native groups as hierarchical, although it was also acceptable to the *pays d'en haut* because kinship systems within many native groups were matrilineal, in which the roles played by fathers with respect to children was often doting and subordinate to the role played by mothers and uncles.[161]

In 1701, the Iroquois ceded the upper country to the French, and in the absence of the Iroquois threat, the *raison d'etre* of the French-led alliance dissipated, leading some of the *pays d'en haut* to discover new grievances amongst each other and to move back east into the Ohio country.[162] Later in mid-century at the start of the Seven Years War, the French had tried to mobilize elements of the *pays d'en haut* against the British, and to do so at times they attempted to harness the potential of Indian rituals and ceremonies. For instance, in Fort Duquesne in 1775, the French officer Daniel de Beaujeu, in attempting to mobilize a contingent of Chippewas, Hurons, Ottowas, Shawnees, Mingos, and Potawatomis "began the Warsong and all the Indian Nations Immediately joined him except the Poutiawatomis of the Narrows [Detroit] who were silent."[163] The Potawatomis did in fact march with de Beaujeu the next day, and following the conclusion of the Seven Years War, the *pays d'en haut* figured prominently in Pontiac's rebellion. In fact, the rebellion was led by one of their own, as Pontiac hailed from the Ottawa tribe. After the British victory over the French, the new overlords of North America decided to dispense with the glue that maintained the French-led alliance with the *pays d'en haut*, particularly the diplomatic protocols of the middle ground and the accompanying provision of gifts and supplies. As previously noted, Lord Jeffrey Amherst, the commander of all British forces in North America, held a dim view of native groups, and facing budgetary pressure from London following the war with the French, he drastically cut the budget for gifts and supplies to native groups (which included blankets, tools, and guns different native groups required for subsistence).[164]

Pontiac's rebellion among the *pays d'en haut* expressed frustration that the British had not effectively replaced the French as Onontio, and the rebels hoped to provoke a French return. Another factor may have been a desire to return to an idyllic pre-Columbian world untainted by whites. Following the Seven Years War, a number of misfortunes plagued the native population east of the Mississippi, including the onrush of white settlers onto their territory, the refusal of the British to adhere to the requirements of the middle ground, the disappearance of game from the frontier as a result of population pressures and overhunting, and the spread of disease epidemics in various native communities.[165]

Elements of the Indian population, including *pays d'en haut*, fell under the spell of various mystics and authority figures calling for separation from whites and abandonment of economic interactions with them and the reemphasis on native traditions and practices. Referred to in the literature as "nativists," one such figure was the Delaware prophet Neolin. In the early 1760s Neolin preached that misfortune had befallen native groups as a result of their complete dependence on European goods, including rum, which fostered vices such as widespread drunkenness. These vices hampered hunting and agriculture and undermined respect for traditional authority structures and figures within native society. Additionally, Neolin faulted native groups for their neglect of ritual and ceremonies, which from a utilitarian perspective, was a form of harnessing spiritual power by appeasing the multitude of supernatural spirits that inhabited the native pantheon of cosmic entities.[166, 167]

Blending elements of both Christianity and native spiritualism, Neolin preached a liberation theology that emphasized separation from whites and their civilization, which found appeal among a variety of constituent groups within the *pays d'en haut*, including the Potawatomis, Wyandots, Miamis, and Ottawas.[168] Dowd suggested that Pontiac was a genuine believer of the nativist message espoused by Neolin, rather than a cynical "political entrepreneur" who was too sophisticated to believe the prophet's message but who nonetheless exploited the sensitivities roused by nativism to mobilize forces against the British.[169] In any case, the nativist message was that salvation, redemption, and uplift lay in the hands of the Indians, through a rejection of the further expansion of both Euro-American settlers and civilization, and the restoration of Indian rituals and ceremonies.

Following the Seven Years War, the English assumed the mantle of Onontio in the early 1760s.[170, 171] During the revolution itself, White noted that those members of the *pays d'en haut* who were most impacted by white settlement, such as tribes that depended on hunting grounds in Kentucky, took up actual fighting, while those less affected participated less.[172] As already noted, the war along the frontier was a brutal race war with immense territorial stakes, and Indian raids wreaked havoc in the backcountry. For their part, settlers did not simply limit their attacks to Indian enemies that perpetrated attacks and atrocities. They also attacked natives who warned settlers of raids and scouted for their military missions, as well as women, children, and Christian pacifists as they prayed, leading White to conclude that murder was the de facto American Indian policy.[173]

By the end of the 1770s, much of the Indian population along the frontier from the Great Lakes to Florida was arrayed against the Patriots,[174] and the British devised plans to mobilize this population to subdue the frontier. Henry Hamilton, the British commander at Vincennes, planned to have the Chickasaws, Cherokees, Shawnees, and Delaware attack in the south; the Mingo, Miami, Wyandot, and Seneca attack from the north; and the Potawatomi, Ottawa, Huron, and Chippewa to attack Kaskaskia. Hamilton hoped to bring about "the greatest Number of Savages on the Frontiers that has ever been known."[175] The capture of Vincennes by George Rogers Clark derailed this ambitious plan, but nonetheless late in the war, native forces, with the assistance of British supplies, put significant pressure along the frontier. Dowd noted:

> Such cooperation, backed by English supplies, made possible the remarkable success of Indians in the American Revolution. While the king's armies surrendered in the East in 1781, Native Americans everywhere north of the Creek country had regained considerable power, if not the military initiative. Between Yorktown and the fall of 1782, Anglo-American settlements from Tennessee through Pennsylvania saw some of the heaviest raiding ever.[176]

British support, though, eventually ended. Calloway noted that:

> Then, just as it seemed the Shawnees were winning the war, Britain snatched them from the jaws of victory. British officers and agents were suddenly urging the chiefs to restrain their warriors and tried to sell them the Peace of Paris as offering a new era of peace with the Americans. When hunters from the Chillicothe lost stock to American horse thieves in the summer of 1783, Major [Arent] De Peyster at Detroit regretted their loss but pointed out "the times are very Critical...the World wants to be at Peace & its time they should be so. [If the Shawnees took action,] it must be an affair of your own, as your Father can take no part in it.[177]

Conflict over territory between native groups and the newly independent United States continued after the revolution and, indeed, well into the nineteenth century.

THE FRENCH AND INDIAN WAR

From Victory to Crisis

If we were to search for the root of the Patriot insurgency, we could conceivably retreat all the way back to the late fifteenth century and the European discovery of North America. More recently, we might light upon the English Civil War or the founding of the English colonies in America. However, from these foundations, a wide variety of outcomes might have proceeded and not necessarily culminated in the formation of a sustained insurgency against the mother country. Instead, we must look at what historians agree was the clear catalyst that led to the American Revolution: the French and Indian War of 1754-1763.

As an extension of the greater conflict known as the Seven Years' War, the French and Indian War unfolded in British and French North America. The French had established a successful enclave in Canada centered on the cities of Quebec and Montreal, and from this base, they were able to project the threat of military power on the Great Lakes and the British colonies of New England. Key to their success was the intricate alliances that they achieved with the Indians of the *pays d'en haut*—the regions around the Great Lakes. Through a diplomatic system based on gift giving, French governors were able to partner with the Indians against British interests. This union gave the French the ability to menace the entire western frontier of the English colonies through their proxies, the various Indian tribes, who excelled at raiding and plundering.

The British response to this threat was a series of attempts to conduct conventional land and sea campaigns against French Canada. To accomplish this, however, the British military commanders and governors tried to coerce cooperation from the American colonists through orders and decrees. Expressing their hubris and cultural arrogance, the British leaders believed that they had the right to command both the colonists and the friendly Indian tribes. Instead, their haughty approach resulted in, at best, half-hearted cooperation. British officers observed colonial militias in operations throughout the war and concluded that they were inferior, ill disciplined, unreliable, and no match for regulars in battle. This perception shaped British political and military decisions in the early years of the Revolution.[178]

From 1754 through 1757, the British suffered a series of defeats. The French and their highly effective Indian allies were able to dominate the crucial Lake Champlain Valley and the Great Lakes. Indian raids on the colonies' western frontier forced British commanders to disperse their limited combat power in a largely vain attempt to stop the raids. The one significant success for British arms came with the conquest of Acadia in 1755, but a disastrous defeat at the Battle of Fort William Henry in 1757 led to the collapse of the British government and the rise of Whig Member of Parliament, William Pitt, First Earl of Chatham.

Pitt (known as Pitt the Elder to distinguish him from his son, who also served as British Prime Minister) was a man of great vision who fundamentally changed the direction of the British war effort. Pitt decided to make French Canada the main effort in the war, and he was determined to eradicate the French from North America. To that end, he borrowed enormous sums of money to subsidize his Prussian ally and to fund the naval and military campaigns he envisioned. He promoted younger, abler generals and admirals, and in the course of two years, 1758-1759, the war swung decisively to England's advantage. At the Battle of Quiberon Bay, the Royal Navy smashed the French Navy, preventing Paris from reinforcing Canada. In short order, British regulars, ably assisted by colonial militias, defeated the French, capturing Montreal and Quebec. Eventually, the French in Canada surrendered, and the British took over the provinces.

The cost of the British victory was enormous. Subsidizing their Prussian allies as well as funding expeditions in Canada, the Caribbean, West Africa, and India had nearly doubled the national debt from 74.5 million pounds in 1755 to 133.25 million pounds in 1763. When William Pitt left office under pressure from the new king, George III, the new government struggled to pay off the debt or at least keep up with the ever-growing interest payments. With Parliament strapped for cash, the members of Parliament concluded—logically, from their perspective—that the American colonies should bear more of the burden for paying off the debt. After all, they reasoned, the removal of the French from Canada was much to the advantage of the Americans. It was only right that they should contribute.

The Americans, however, did not see it that way. First, they felt that they had contributed by sending militias to support the British regulars. They provided much-needed logistical support during the various campaigns, and they believed that England herself benefitted more from the removal of the French than the colonies. When the British government began to levy taxes on the colonists, they reacted by insisting that Parliament had no constitutional power to tax the colonies because they were not represented in Parliament. This issue became the central point of contention that led to war.

Two factors proceeding from the French and Indian War led directly to the formation of the Patriot insurgency. First, the eradication of the age-old French threat from Canada removed the American colonists' dependence on the mother country for protection. Second, the intractable British debt led a succession of ministers to attempt to squeeze revenue from the colonies in ways that were questionable on constitutional grounds. Thus, the great victory wrought by William Pitt was also the origin of a crisis that would lead to the start of the collapse of British imperialism.

ENDNOTES

1. Robert Middlekauff, *The Glorious Cause: the American Revolution, 1763-1789* (Oxford: Oxford University Press, 1982), 39.
2. Ibid., 40.
3. Ibid.
4. Schmidt, Ethan A. *Native Americans in the Revolution: How the War Divided, Devastated, and Transformed the Early American Indian World* (Santa Barbara, CA: Praeger, 2014), 1.
5. Edward Countryman, "Indians, the Colonial Order, and the Social Significance of the American Revolution," *The William and Mary Quarterly*, Vol. 53, No. 2 (April 1996): 354.
6. Alan Taylor, *American Colonies: The Settling of North America* (New York: Penguin Books, 2001), 93.
7. Ibid., 97; J. Russell Snapp, *John Stuart and the Struggle for Empire on the Southern Frontier* (Baton Rouge, Louisiana State University Press, 1996), 13.
8. Taylor, *American Colonies*, 98.
9. Barbara Graymont, *The Iroquois in the American Revolution* (Syracuse, Syracuse University Press, 1972), 23-24.
10. Ibid., 94.
11. Ibid., 17-18.
12. It is hard to overstate the magnitude of the demographic calamity brought about by the transference of Old World diseases to the Americas. As noted by Taylor, "In any given locale, the first wave of epidemics afflicted almost every Indian. Within a decade of contact, about half the natives died from the new diseases. Repeated and diverse epidemics provided little opportunity for native populations to recover by reproduction. After about fifty years of contact, successive epidemics reduced a native group to about a tenth of its precontact numbers. Some especially ravaged peoples lost their autonomous identity, as the few survivors joined a neighboring group. Consequently, the Indian nations ("tribes") of colonial history represent a subset of the many groups that had existed before the great epidemics...Recognizing this demographic catastrophe, recent scholars have dramatically revised upwards their estimates of the pre-Columbian population in the Americas. Because the natives lacked statistical records (and their first conquerors rarely kept any), all calculations of the contact populations are highly speculative...Early in the twentieth century, most scholars were "low counters," who estimated native numbers in 1492 at only about 10 million in all of the Americas, including about one million north of the mouth of the Rio Grande. More recent scholars, the "high counters," ... doubled the estimated population of the pre-Columbian Americas to twenty million. Some insist upon 100 million or more. Narrowing their view to just the lands north of the Rio Grande, the revisionists claim that the future United States and Canada together contained at least two and perhaps ten million people in 1492. Most scholars now gravitate to the middle of that range: about fifty million Indians in the two American continents, with about five million of them living north of Mexico."
13. Taylor, *American Colonies*, 39-40.
14. Schmidt, *Native Americans in the Revolution*, 56-57, 59, 60.
15. Snapp suggested that the desire to dispossess Native groups of their lands may have been more prominent by the 1770s than in earlier periods. He noted that "On the eve of the Revolutionary War, and even more so during and after it, many southerners concluded that the Indians had outlived their usefulness. Since the first European settlements in southeastern North America, whites at all levels of society, from the early Augusta traders

to imperial officials, had regarded natives as useful trade partners and military allies. By the early 1770s, however, many of the increasingly numerous Anglo-Americans saw them primarily as an obstacle to the rapid expansion of prosperous white settlements."

[16] Snapp, *John Stuart and the Struggle for Empire on the Southern Frontier*, 216.

[17] Taylor, *American Colonies*, 135.

[18] Graymont, *The Iroquois in the American Revolution*, 89.

[19] Schmidt, *Native Americans in the Revolution*, 24.

[20] Taylor, *American Colonies*, 421.

[21] Ibid., 18.

[22] Snapp, *John Stuart and the Struggle for Empire on the Southern Frontier*, 70.

[23] Taylor noted that British taxpayers, rather than colonial subjects in North America, were largely shouldering the expenses of empire. Specifically, in 1763 the former were taxed 26 shillings per person, while those in the colonies were paying one shilling per person.

[24] Taylor, *American Colonies*, 438-39.

[25] Schmidt noted that "In particular, the Pontiac War taught the British government that it possessed neither the financial nor the military strength to immediately quell another such rebellion. In light of this, King George III and many of his ministers recognized the dire need to stem westward settlement as a way of minimizing the chances of another major conflict between whites and Indians."

[26] Schmidt, *Native Americans in the Revolution*, 12, 23, 79.

[27] In addition to financial concerns, balance of power considerations in North America also factored into British thinking. In particular, John Stuart was concerned that Anglo-American policies and attitudes towards Native Americans might drive them into an alliance with the Spanish. Snapp noted that "In 1777, for example, he [Stuart] noted the distinct difference between Spanish policy toward the Indians in Louisiana and prevailing Anglo-American attitudes. The Spanish, he reported to General Howe, hoped to increase their power in the region by enticing as many Indians as possible to move to their side of the Mississippi. Anglo-American behavior, however, indicated a desire 'to wish them [the Indians] gone for the sake of their Land'."

[28] Snapp, *John Stuart and the Struggle for Empire on the Southern Frontier*, 216.

[29] The perception of inferiority appears to have been mutual. Calloway noted that "Indians generally do not appear to have possessed any great veneration for either white men or their civilization. Rather, they regarded the whites' way of life and their claims to superiority with disdain and the white men themselves as inferior. Initial contacts might see the Indians viewing European newcomers with awe, but the impression soon wore off, as fur traders, with their repertoire of poses and techniques for maintaining Indian respect, knew only too well. Europeans were at a disadvantage from the beginning of contact because they could hardly match the Indian in the Indian world. As trader Peter Grant explained about the Sauteux, or Chippewas: 'Though they acknowledge the superiority of our arts and manufacturers, and their own incapacity to imitate us, yet, as a people, they think us far inferior to themselves. They pity our want of skill in hunting and our incapacity of travelling through their immense forests without guides or food…The highest compliment which they bestow on a white man is that he is in every respect like one of themselves, but no man can aspire to that honor who has not a tolerable knowledge of their language and customs.' To such Indians, British intruders bore a strange appearance, wore impractical and uncomfortable clothing, spoke in unintelligible dialects, and indulged in incomprehensible and even antisocial behavior."

30. Snapp, *John Stuart and the Struggle for Empire on the Southern Frontier*, 68; Colin G. Calloway, *Crown and Calumet: British-Indian Relations, 1783-1815* (Norman: University of Oklahoma Press, 1987), 102-103.

31. Ibid., 70

32. Ibid., 76.

33. Schmidt suggested that what most rankled frontier settlers was that Crown actions seemed to imply an equivalence between colonists and Indians: "The very idea that British government officials would take the part of Indians against whites incensed backcountry settlers and unscrupulous traders alike. Herein lies one of the principal disputes that underlay the imperial crisis in the south (and elsewhere in colonial America), which ultimately brought on the American Revolution there. The British government remained willing to conceive of Native Americans as subjects of the crown, similar to colonists. American colonists, in this case those living in the southern colonies, refused to see Indians as fellow subjects. Instead, they viewed them as obstacles in the way of their dreams of land ownership and trading wealth."

34. Snapp, *John Stuart and the Struggle for Empire on the Southern Frontier*, 3; Schmidt, *Native Americans in the Revolution*, 33.

35. Taylor, *American Colonies*, 103.

36. In addition to helping restore a tribe's power and enhance their own status, Iroquois warriors engaged in mourning wars to enable the families of those killed in war to properly grieve, which traditionally was best achieved in Iroquoian culture through the replacement of the killed family member with a suitable war captive to take the place of the deceased family member.

37. Taylor, *American Colonies*, 102-103.

38. Fred Anderson, *Crucible of War: The Seven Years' War and the Fate of Empire in British North America, 1754 – 1766* (New York, Vintage Books: 2001), 14; Taylor, *American Colonies*, 104.

39. Anderson, *Crucible of War*, 14.

40. Schmidt, *Native Americans in the Revolution*, 54.

41. Taylor, *American Colonies*, 262.

42. David Levinson, "An Explanation for the Oneida-Colonist Alliance in the American Revolution," *Ethnohistory* 23, no. 3 (Summer 1976): 267.

43. Caitlin A. Fitz, "'Suspected on Both Sides': Little Abraham, Iroquois Neutrality, and the American Revolution," *Journal of the Early Republic*, 27 (Fall 2008): 299.

44. Levinson, "An Explanation for the Oneida-Colonist Alliance in the American Revolution," 272.

45. Schmidt, *Native Americans in the Revolution*, 129-130.

46. Ibid., 132, 136.

47. Colin G. Calloway, *The American Revolution in Indian Country: Crisis and Diversity in Native American Communities* (Cambridge, Cambridge University Press: 1995), 132-133.

48. Following the British defeat of France in the Seven Years War, the British eventually assumed the cultural role of "father" to various Indian tribes.

49. Jon William Parmenter, "Pontiac's War: Forging New Links in the Anglo-Iroquois Covenant Chain, 1758-1766," *Ethnohistory* 44, no. 4 (Autumn 199): 636. For a discussion of how the French assumed the cultural role of "father" in their relationship with the *pays d'en haute*, see the discussion on that native group.

50. Graymont, *The Iroquois in the American Revolution*, 58.

51 Levinson, "An Explanation for the Oneida-Colonist Alliance in the American Revolution," 265.

52 Schmidt, *Native Americans in the Revolution*, 127.

53 Levinson, "An Explanation for the Oneida-Colonist Alliance in the American Revolution," 266, 278, 280.

54 Ibid., 280-283.

55 Such fears may explain the Oneida reaction in 1776 to pressure from other Iroquois nations to support the British: "Brother, we dread the consequences." Tiro suggested that another Oneida motivation for supporting the colonists was vulnerability, given concern that fighting their immediate neighbors would only hasten their dispossession.

56 Levinson, "An Explanation for the Oneida-Colonist Alliance in the American Revolution," 280-281; Karim M. Tiro, "A 'Civil War? Rethinking Iroquois Participation in the American Revolution," *Explorations in Early American Culture* 4 (2000): 150.

57 Levinson, "An Explanation for the Oneida-Colonist Alliance in the American Revolution," 270.

58 The Oneida also informed the Committee of Safety of efforts by other Iroquois tribes to sway them to neutrality. For instance, in May 1775, they provided the Committee with a letter authored by the Mohawks urging the Oneida not to aid the patriots.

59 Levinson, "An Explanation for the Oneida-Colonist Alliance in the American Revolution," 271.

60 Tiro noted that while different factions of the Iroquois were indeed arrayed against each other during the revolution, on a deeper level they still identified with one another given Euro-American expansionism and hostility towards native groups following the defeat of France in the Seven Years War. This in turn led to a limit in the level of violence they inflicted on each other. He noted that "In the woods, beyond the sight of Europeans, pro-British and pro-Patriot Iroquois met and exchanged information, and the courtesy was reciprocated. While Indians on both sides devastated one another's villages, they spared one another's lives. Indeed, it is extremely telling that so few Iroquois died at the hands of other Iroquois. All told, Indian-on-Indian violence during the war appears muted when compared with both Indian-on-white violence and white-on-Indian violence." Additionally, Tiro noted the ambivalent participation of the Oneida in General John Sullivan's 1779 campaign against Iroquoia, a hesitancy which was anticipated by General Washington. While Washington suspected the reluctance of the Oneida and Tuscarora to participate in operations against the Confederacy, he also knew they possessed intimate knowledge of Iroquois territory. Hence he issued instructions to "endeavour to get [this information] from the Friendly Oneidas in such a manner as not to give them any suspicions of the real design."

61 Schmidt, *Native Americans in the Revolution*, 134; Tiro, "A 'Civil War?, 148, 149, 157,158.

62 Calloway, *The American Revolution in Indian Country*, 86.

63 Ibid. Schmidt, *Native Americans in the Revolution*, 51

64 Interestingly, the Mohawks warned the Stockbridge against taking up arms during the revolution, stating "If any ill Consequences should follow, you must conclude you have brought it upon yourselves, and that it is your own fault."

65 Calloway, *The American Revolution in Indian Country*, 94.

66 Ibid., 96.

67 Ibid., 97.

68 Ibid., 93.

69 Schmidt, *Native Americans in the Revolution*, 122.
70 Ibid., 90-92.
71 Ibid., 52.
72 Graymont, *The Iroquois in the American Revolution*, 6.
73 Taylor, *American Colonies: The Settling of North America*, 434.
74 Schmidt, *Native Americans in the Revolution*, xiii.
75 Calloway, *The American Revolution in Indian Country*, 182.
76 Ibid., 189.
77 Ibid., 189, 201.
78 Ibid., 191.
79 Ibid., 195.
80 Schmidt, *Native Americans in the Revolution*, 90.
81 Ibid., 91.
82 Ibid.
83 Calloway, *The American Revolution in Indian Country*, 201.
84 Schmidt, *Native Americans in the Revolution*, 91.
85 Ibid., 93.
86 Ibid., 94.
87 Ibid., 29.
88 A persistent complaint of tribes in the southern colonies was that agents licensed by various colonies to engage in the Indian trade often engaged in unscrupulous trade practices to the detriment of the tribes. Such practices consisted of the excessive selling of rum and other liquor, and the encouragement of debt accumulation to the point where debts could only be discharged through the sale of land.
89 Schmidt, *Native Americans in the Revolution*, 34-35.
90 Ibid., 95.
91 Agreements with native tribes were often solidified with the provision of gifts. Taylor noted that "In the native world there was no mediation, no meaningful public action, without the delivery of gifts. Words were pointless and no agreement was binding unless accompanied by presents. The greater the ceremony and the offerings involved, the more serious and binding the agreement."
92 Taylor, *American Colonies*, 380.
93 Ibid., 97.
94 Ibid., 98.
95 Ibid., 99.
96 Ibid., 100.
97 Ibid., 100-101.
98 Ibid., 102.
99 Ibid., 103.
100 Calloway, *The American Revolution in Indian Country*, 246.
101 Ibid., 249.
102 Schmidt, *Native Americans in the Revolution*, 98.

[103] Calloway, *The American Revolution in Indian Country*, 250.

[104] Ibid., 268.

[105] Ibid., 266.

[106] Ibid., 160.

[107] Schmidt, *Native Americans in the Revolution*, 81.

[108] Ibid., 82.

[109] Lord Dunmore's war represented perhaps the epitome of the contradiction inherent in British Indian policy. As the appointed royal Governor of Virginia, John Murray, the Fourth Earl of Dunmore, represented the executive authority of the empire in Virginia, and Schmidt suggested that one motivation for the war with the Shawnees was to divert public opinion away from the revolutionary rhetoric of Patrick Henry. Another likely motive was the desire to acquire Indian territory further west, which Schmidt noted was in violation of standing policy of His Majesty's government, specifically the Proclamation of 1763.

[110] Schmidt, *Native Americans in the Revolution*, 83.

[111] The Shawnee policy of dispersal appears to have been a common practice up to this point, and Schmidt noted that this policy was based on the Shawnee preference to disperse to other territories rather than confront Europeans as a unitary group congregated in a specific location. Calloway suggested that this preference was based on a cultural conservatism within the group, reflecting a desire to keep Shawnee culture untainted by elements of European civilization, such as Christianity.

[112] Schmidt, *Native Americans in the Revolution*, 72, 163.

[113] Ibid..

[114] One notable feature of the increased British presence that did not escape the notice of the Shawnees was the massive Fort Pitt, built on the ruins of French-built Fort Duquesne at the juncture of the Monongahela and Alleghany rivers. Fort Pitt was ten times the size of its French predecessor, capable of housing over 1,000 troops and 20 artillery pieces. Anderson noted that the fort represented a "symbol of dominion, an emblem of empire; and…the Ohio Indians were beginning to discern its meanings only too well."

[115] Schmidt, *Native Americans in the Revolution*, 329.

[116] Some Shawnee groups though did continue the practice of migration. In 1773, the Seneca chief Kayashuta noted that 170 Shawnee families moved away from their location along the Scioto river in the Ohio Valley to avoid being "Hemmed in on all Sides by the White People, and then be at their mercy."

[117] Calloway, *The American Revolution in Indian Country*, 161.

[118] Colin G. Calloway, "'We Have Always Been the Frontier': The American Revolution in Shawnee Country," *American Indian Quarterly* 16, no.1 (Winter 1992): 40.

[119] Schmidt, *Native Americans in the Revolution*, 142.

[120] Gregory Evans Dowd, *A Spirited Resistance: The North American Indian Struggle for Unity, 1745–1815* (Baltimore: Johns Hopkins University Press, 1992), 67.

[121] Calloway, *The American Revolution in Indian Country*, 162.

[122] Dowd, *A Spirited Resistance, 1745–1815*, 68.

[123] Schmidt, *Native Americans in the Revolution*, 143.

[124] Dowd, *A Spirited Resistance, 1745–1815*, 67.

[125] Schmidt, *Native Americans in the Revolution*, 144.

126 Dowd, *A Spirited Resistance, 1745–1815*, 75.

127 Calloway, "'We Have Always Been the Frontier'," 42-3.

128 Eric Hinderaker, *Elusive Empires: Constructing Colonialism in the Ohio Valley, 1673 – 1800* (Cambridge: Cambridge University Press, 1997), 220-224; Calloway, "'We Have Always Been the Frontier'," 43.

129 Schmidt, *Native Americans in the Revolution*, 151.

130 Ibid., 152.

131 Earlier in late 1777/early 1778, Clark had petitioned Virginia governor Patrick Henry to authorize the raising of militia forces to attack British outposts in Illinois country, a territory which encompasses modern-day Indiana, Illinois and parts of Ohio and Kentucky. Schmidt noted that, strictly speaking, the Virginia governor did not have the authority to authorize an expedition so far afield, and that it was unlikely that many Virginians were willing to serve in an expedition so far from their homes. Hence Henry issued two sets of instructions: the public authorization called on Clark to defend Kentucky, while the private one, which was kept secret from most of the Virginia legislature, authorized Clark to attack Kaskaskia, in present-day Illinois, and to keep the true destination of his force secret. Schmidt argued that the private instructions offered to Clark had less to do with protecting the frontier from, say, British tyranny, but rather to acquire Indian territory by force for Virginia.

132 Schmidt, *Native Americans in the Revolution*, 145-147.

133 Calloway, "'We Have Always Been the Frontier'," 46.

134 Ibid., 47.

135 In fact, one of the more notorious atrocities in the Ohio Valley occurred following the surrender in Yorktown. In March 1782, an American militia massacred 96 pacifist Indians at the Moravian mission village of Gnadenhutten, in Ohio. The militia members were motivated by a desire to avenge Indian attacks, yet Schmidt noted that they were aware that the Moravian Indians were not involved in attacks on their settlements. One eyewitness painted a gruesome and macabre picture: "In the other house, Judith, an aged and remarkably pious and gentle widow was the first victim. Christina, before mentioned, fell on her knees and begged for life. In vain! In vain! The tigers had again tasted blood. In both houses men, women, and children were bound by ropes in couples, and were thus led like lambs to the slaughter. Most all of them, I heard – for I only saw that part of the butchery which I was compelled to witness – marched cheerfully, and some smilingly to meet their death." See Schmidt, Native Americans in the Revolution, 153. Interestingly, in his summary of the incident 100 years later in the book The Winning of the West, Theodore Roosevelt wrote that "It is impossible not to regret that fate failed to send some strong war party of savages across the path of these inhuman cowards, to inflict on them the punishment they so richly deserved." The former President appears not to have been aware that fate did indeed exact a measure of revenge. Some of the perpetrators participated in a subsequent expedition led by Colonel William Crawford and Colonel David Williamson that saw battle in Sandusky, Ohio against Shawnees and Delaware Indians in June 1782. The expedition was defeated, and those who had participated in the Gnadenhutten massacre were tortured and burned alive.

136 Dowd, *A Spirited Resistance*, 86.

137 In this regard, the conflict in the Ohio Valley bears a strong resemblance to many other conflicts in which local concerns often overshadow the factors which initially motivated a conflict.

Chapter 3. Historical Context

[138] Stathis N. Kalyvas, *The Logic of Violence in Civil War* (New York: Cambridge University Press, 2006): 364-387.

[139] Schmidt, *Native Americans in the Revolution*, 141.

[140] Dowd, *A Spirited Resistance, 1745–1815*, 77.

[141] Ibid., 69-71.

[142] Ibid., 76.

[143] Ibid., 80-81.

[144] Schmidt, *Native Americans in the Revolution*, 36.

[145] Ibid., 37.

[146] Ibid.; Calloway, *The American Revolution in Indian Country*, 221.

[147] Schmidt, *Native Americans in the Revolution*, 105.

[148] Calloway, *The American Revolution in Indian Country*, 224.

[149] Ibid., 224-225.

[150] Ibid., 226.

[151] Schmidt, *Native Americans in the Revolution*, 109.

[152] Taylor, *American Colonies*, 376.

[153] Ibid., 378.

[154] Ibid., 377-378.

[155] Ibid., 377.

[156] Richard White, *The Middle Ground: Indians, Empires, and Republics in the Great Lakes Region, 1650 – 1815* (Cambridge: Cambridge University Press, 1991), 30.

[157] Ibid., 50-53.

[158] Taylor, *American Colonies*, 379.

[159] White, *The Middle Ground*, 36.

[160] Taylor, *American Colonies*, 380.

[161] Ibid.

[162] Ibid., 301.

[163] Dowd, *A Spirited Resistance, 1745–1815*, 25.

[164] Schmidt, *Native Americans in the Revolution*, 18-19.

[165] Ibid., 23.

[166] Hence, various mystics blamed misfortune on the anger of neglected spirits. When the Pennsylvanian Indian agent Conrad Weiser asked a Shawnee in the 1730s why their territory had become unproductive, Weiser stated that "They answered…that the Lord and creator of the world was resolved to destroy the Indians. One of their seers, whom they named, had seen a vision of God, who said to him the following: - You inquire after the cause why game has become scarce, I will tell you. You kill it for the sake of the skins, which you give for strong liquors, and drown your senses, and kill one another, and carry on a dreadful debauchery. Therefore have I driven the wild animals out of the country, for they are mine. If you will do good, and cease from your sins, I will bring them back; if not, I will destroy you from the end of the earth." Schmidt noted that the fact that the Shawnee referenced a single supernatural entity rather than a specific spirit particular to a tribe reflected the pan-Indian nature of the nativist movement sweeping through various Indian groups and the ability of this movement to bridge cultural and spiritual divides separating distinct tribes.

167 Dowd, *A Spirited Resistance*, 1-4, 33-36; Schmidt, *Native Americans in the Revolution*, 22-23.

168 Ibid., 34.

169 Ibid., 35.

170 The Shawnees were cruelly reminded of this fact in July 1783, when one American officer informed a Shawnee that "Your Fathers the English have made Peace with us for themselves, but forgot you their Children, who fought with them, and neglected you like Bastards." See Schmidt, Native Americans in the Revolution, 155. Indeed in the Treaty of Paris following the war, during which there was no direct Indian participation, the British ceded Indian territory to the United States. As noted by Calloway, "After the Peace of Paris, the United States claimed the Indians' lands by right of conquest and proceeded to "give peace" to the tribes in a series of separate treaties, in effect granting the Indians portions of their own lands. In American eyes, the Indians could have no complaints. It seemed only just that former enemies should make some atonement for the atrocities they had perpetrated during the war; indeed, the United States was being generous in not demanding that the tribes remove to Canada. The Indians, however, saw themselves not as defeated subjects of George III but as free peoples. They had suffered no defeat to merit the loss of their lands, nor had they given their consent to the Peace of Paris. The tribes considered their territorial boundaries those set by various pre-Revolutionary treaties to which they had agreed, not by the Peace of Paris to which they had not. In particular, the northern and western tribes adhered to the Treaty of Fort Stanwix of 1768 which had fixed the Ohio River as the boundary between Indian land and white settlement." See Calloway, "Suspicion and Self-Interest," 49.

171 Schmidt, *Native Americans in the Revolution,* 155; Colin G. Calloway, "Suspicion and Self-Interest: The British-Indian Alliance and the Peace of Paris," *The Historian* 48, no. 1 (November 1985): 49.

172 White, *The Middle Ground*, 378.

173 Ibid., 384.

174 Dowd, *A Spirited Resistance, 1745–1815*, 59.

175 Ibid., 57.

176 Ibid., 59.

177 Calloway, "'We Have Always Been the Frontier'," 46-47.

178 Fred Anderson, *Crucible of War: the Seven Years' War and the Fate of Empire in British North America, 1754-1766* (New York: Vintage Books, 2001), 286-296.

CHAPTER 4.
SOCIOECONOMIC CONDITIONS

The distinctions between Virginians, Pennsylvanians, New Yorkers, and New Englanders are no more. I Am Not A Virginian, But An American!

—Patrick Henry

POPULATION

The population of the English colonies at the start of the American Revolution featured a mix of descendants of European immigrants, black slaves, and Native Americans. The population in 1775, excluding Native Americans, was approximately 2,500,000.[1] Most were of English descent. About 20 percent of that number consisted of enslaved African blacks (i.e., roughly 500,000). The Scotch-Irish made up the next largest group—immigrants from Ireland who fled the religious persecution that followed the Glorious Revolution. German Protestants likewise immigrated mostly to Pennsylvania, but small numbers were found throughout the thirteen colonies. The rest of the immigrant population included representatives from nearly every European country (see Figure 4-1).

The most important demographic dynamic in the eighteenth century was the extraordinarily rapid growth of the population. From 1700 to 1750, the colonial population expanded tenfold. This influx of people contributed to economic growth, but it also served to loosen the bonds between colonial authorities and the mother country. A variety of cultures came together in the colonies, and a substantial portion of the population felt no great love for King and Parliament.

Social Classes

Colonial society included clear class distinctions, and the most defining factor in determining class was land. The social and political elites were generally large land owners who had possession of vast tracts from which they extracted leases, quit-rents, and other fees. Doctors, lawyers, wealthy merchants, and bankers also populated the top strata. These gentry provided much of the political and military leadership for the Patriot insurgency.

America's pre-industrial society did not have a modern urban middle class. Instead, merchants and related businessmen usually dominated the few urban centers, and some in this category became quite wealthy. This so-called "middling sort" included what would today be thought of as middle class, although the concept of a large, stable, and politically dominant middle class was unheard of at the time. This class included tradesmen, craftsmen, teachers, lesser merchants, and other urban workers. Middling sorts often owned property, and males could vote.

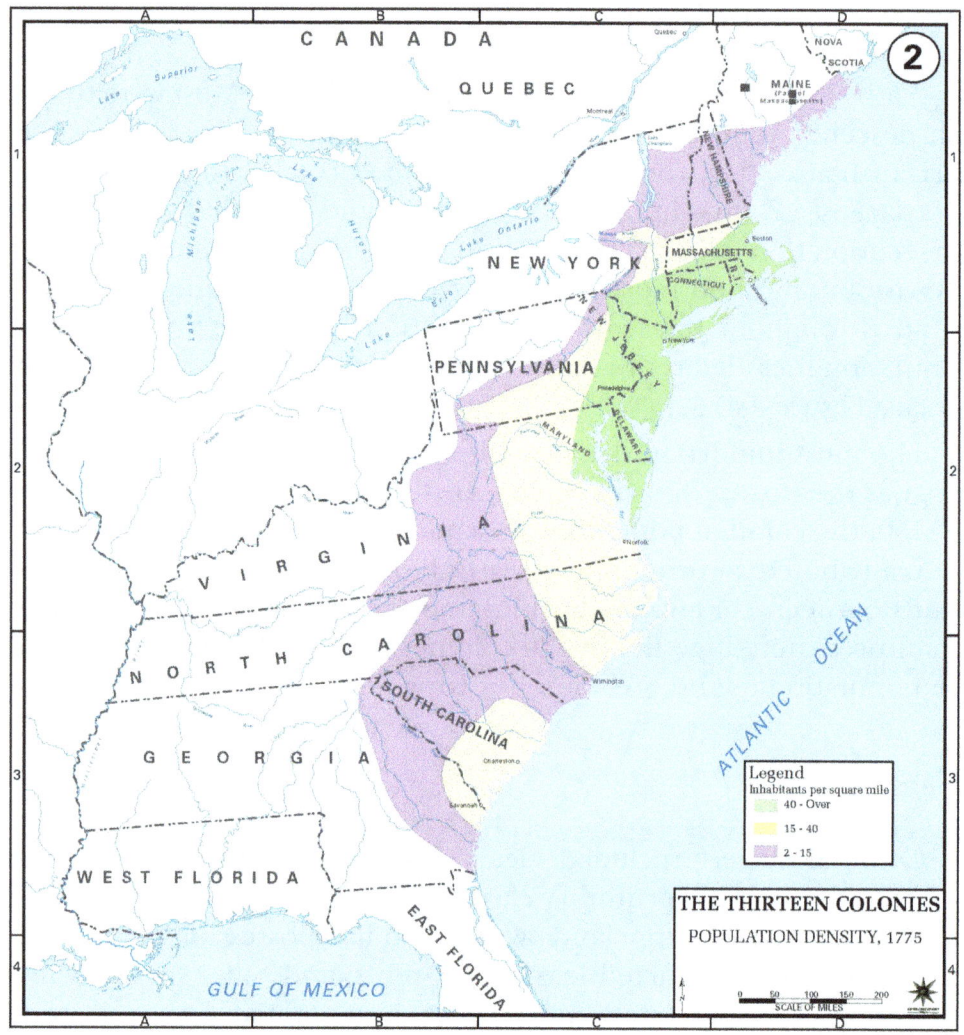

Figure 4-1: Population density, 1775.

Much of the population was involved in farming, and colonial American included a large class of small farm owners. Socially, they fell between the lower class and the middling class. Most operated family-owned farms and supplemented family labor with a few slaves or hired workers. Besides farming for family consumption, they often produced cash crops—tobacco, cotton, wheat, and corn. The difficulty of storing food crops and transporting them to market led many farmers into whiskey and beer production.

There were also a substantial number of impoverished whites both in the cities and on farms. Urban poor were a considerable burden to municipal authorities, who had few institutional solutions to house, feed, and provide medical care to the unemployed. Rural poor often worked as indentured servants or low-wage workers on farms, many hoping to one day own their own land.

Enslaved blacks likewise made up a substantial portion of the population. Unlike their brothers and sisters enslaved in the Caribbean sugar islands, American blacks enjoyed healthier environmental conditions and so tended to live longer and produce more children. Slaves fell into three general categories: enslaved artisans, household servants, and field hands. Household servants had typically better working conditions than field hands. Enslaved artisans were the fastest growing category of the enslaved population in the mid-eighteenth century. The black population grew during the Revolution—a result of slaves escaping and running to freedom or, in some cases, being manumitted by their owners. They provided skilled and unskilled labor, and some owned property; however, they did not enjoy the same civil rights as whites.

Religion

The religious context of the Patriot insurgency was a crucial factor in sustaining the rebellion and in keeping the insurgents from falling into factionalism. Religious motivation lay behind much of the early colonization of the Americas, and by the time of the Revolution, there were many Christian denominations vying for the loyalty of the colonists. Nine of the colonies had an established church—in other words, one supported by public taxes. This institutional arrangement was a holdover from the practice in England, but the most fervently religious in America emerged outside of the established churches. Indeed, this fact accounted for much of the dissatisfaction that some Patriots felt.

New England was home to Congregationalists, descendants of the original Puritan founders. They maintained their position of power within society, despite growing numbers of Anglicans, Quakers,

Baptists, and others who continually pushed back against having to pay taxes to the Congregational establishment.

In the Middle Colonies, a greater degree of religious diversity prevailed. Although this would eventually give rise to the American tradition of religious tolerance, conflict was a constant feature of the pre-Revolution era. Presbyterians, Quakers, Catholics, and Anglicans each had their constituencies and strived to win over converts and special privileges from the government.

In the Southern Colonies, Anglicanism had its strongest hold, and churches there remained in control of public tax support. However, as in the north, other religious groups—the Baptists in particular—pushed back against official religion through non-attendance (in favor of their own meetings) and growing complaints about believers having to pay taxes in support of an official church that they considered sinful.

The most remarkable feature of American religion—particularly after the Great Awakening of 1740-1741—was its ability to unite the colonists. The factional and denominational rifts among them were real, but the similarities in religious outlook among nearly all colonists was powerful. In America, laymen typically exerted much greater power in churches than in Europe. Ministers, consequently, wielded much less official authority. Believers focused more on the worth of individual experiences with God than on staid religious ritual. Enlightenment philosophy led American Christians to question everything and to put value on their own interpretations of spiritual truth. Their proclivity for religious separatism added to revolutionary fervor when the break with England finally came.

ENDNOTES

[1] Robert Middlekauff, *The Glorious Cause: The American Revolution, 1763-1789*, (Oxford, United Kingdom: Oxford University Press, 2007), 32.

CHAPTER 5.
GOVERNMENT AND POLITICS

[M]ay it therefore please your most excellent Majesty, that it may be declared. . . . That the said colonies and plantations in America have been, are, and of right ought to be, subordinate unto, and dependent upon the imperial crown and parliament of Great Britain.

—The Declaratory Act, March 18, 1766

BRITISH NORTH AMERICA

In chapter 3, we examined the history of the thirteen British colonies. In this chapter, we turn to examining how Great Britain ruled—or attempted to rule—those colonies within the context of the empire. In organizing discussion of various British acts of Parliament, we distinguish between expressions of British imperial policy on the one hand, with acts in response to the Patriot rebellion on the other. The former occurred as Great Britain attempted to correct budget deficits and regulate trade within the empire. The latter constituted government countermeasures aimed at quelling the popular revolt that emerged. Thus, the Currency, Sugar, and Stamp Acts resulted from economic policy decisions. The Declaratory and Coercive Acts were responses to Patriot provocations. We will examine the latter category more closely in chapter 8.

BRITISH WEST INDIES

Jamaica and the Sugar Colonies

In the early seventeenth century, England established plantation colonies in the West Indies, the chief of which was the precious Jamaica. To produce the coveted sugar and related products (e.g., rum and molasses) the planters imported thousands of African slaves and had to maintain the flow of the enslaved workers because the climate, diseases, and harsh working conditions routinely killed many slaves. The sugar islands were Britain's most valued possessions in the western hemisphere, but they were linked to the North American colonies in a vital trade relationship. The American colonies produced foodstuffs for the Caribbean slaves, and they also provided a ready market for sugar, rum, and molasses.

As with the French colony of Saint-Domingue, the restive slave population eventually revolted in parts of Jamaica. Escaped slaves retreated to the island's rough interior mountains, from where they maintained their independence. Called "maroons," the former slaves' determination combined with impenetrable mountains thwarted the British attempts to bring them to heel. Instead, the plantation owners enacted a tense truce with the maroons, even paying them to return more recently escaped slaves.

During the American Revolution, Great Britain continued to value the sugar islands and worried that French intervention in America might threaten English holdings in the Caribbean (as indeed it did). Both the Royal Navy and the French Navy found themselves distracted by having to defend their interests there while simultaneously having to attend to the Patriot rebellion to the north.

British Policy and the American Colonies

Following the French and Indian War, Great Britain faced a debt crisis. Possible strategies for reducing that debt included direct taxation on domestic populations, imposition of tariffs, redistribution of government expenditures, and taxation on the American colonies. It was inevitable that Parliament chose the last of these options after 1763 because vested interests among members of Parliament precluded the other possibilities. There was a general feeling among Englishmen at home that they had done their share and that the Americans—whom the victory in war benefited the most—should take on the burden of financing the debt. Because the American colonies had no representatives in Parliament, it was an easy matter for the gentlemen in England to agree.

The Sugar Act of 1764

The first expression of the strategy was the Sugar Act of April 1764. This initial legislation attempted to put an end to the tradition of "salutary neglect" in which British authorities avoided a strict enforcement of trade regulations to keep the Americans quiet. Under financial pressure, the Grenville ministry had to wring taxes from the obstreperous colonists, and the Sugar Act—while reducing the tariff on imported molasses—aimed at actually collecting the revenue. Although the expected income from the tariff would amount to a paltry 79,000 pounds per year—hardly enough to make a dent on British debt—it would help defray the cost of British troops in America.

The colonists objected both to the presence of the troops and to the newly enforced tariff that would support them. Former lax enforcement of the 1733 Molasses Act allowed the colonists to smuggle molasses from French, Dutch, and Spanish islands, which was much cheaper than English molasses. At the insistence of the English planters, Parliament sought to end the smuggling by heavily taxing the competing imports and strictly patrolling for violators.

The other aspect of the legislation that colonists found intolerable was the stipulation that violations of the import and export rules would be adjudicated in admiralty courts in another colony. Previously such criminal activity would be handled in local courts among juries drawn from the colonists. Typically, local courts were sympathetic to the smugglers and often exonerated them or gave them token punishments. Operating without juries, the admiralty courts would take a much sterner view, and they seemed to violate a legal principle of the right to a trial by jury.

Merchants and those engaged in port activities felt the burden most. Samuel Adams gave voice to what some felt was a constitutional issue:

> For if our Trade may be taxed why not our Lands? Why not the Produce of our Lands & every thing we possess or make use of? This we apprehend annihilates our Charter Right to govern & tax ourselves – It strikes our British Privileges, which as we have never forfeited them, we hold in common with our Fellow Subjects who are Natives of Britain: If Taxes are laid upon us in any shape without our having a legal Representation where they are laid, are we not reduced from the Character of free Subjects to the miserable State of tributary Slaves?[1]

In general, however, the public response to the Sugar Act was muted enough to convince the Grenville ministry and Parliament that American colonists would accept their authority in trade and tax policy.

The Currency Act of 1764

In the same year as the Sugar Act, Parliament responded to requests from British merchants to protect them from depreciating colonial paper money by passing the Currency Act. The legislation required colonists to pay public and private debt only with specie, not with paper money. The problem was that gold and silver coins were rare and becoming rarer each year. The mercantilist policies of England and the rest of Europe squeezed the already tight supply in the colonies, with the result that merchants and others could no longer afford to pay their debts. The economic depression that descended on America in the wake of the French and Indian War exacerbated the problem. As with the Sugar Act, the public response to the Currency Act was of small enough scale that Parliament felt they got away with it. However, the growing debt issue contributed a major impulse for revolution among the Patriots.

ENDNOTES

[1] Draper, Theodore, *A Struggle For Power: The American Revolution.* (New York: Vintage, 1997), 219.

PART II.
STRUCTURE AND DYNAMICS OF THE INSURGENCY

CHAPTER 6.
PATRIOT INSURGENCY

We must all hang together, or assuredly we shall all hang separately.

—Benjamin Franklin

Chapter 6. Patriot Insurgency

In the wake of the French and Indian War, the American colonists were experiencing a transitional period in which they were attempting—though not collectively—to restore the economic leeway and political freedom within the British Empire that they had previously enjoyed. At the same moment, however, the resplendent Empire, glorying in its unparalleled global power, yet wallowing in debilitating debt, aimed to rein in the American colonies and impose on them a new institutional inferiority within the imperial system. A clash was looking more likely.

THE COURSE OF THE PATRIOT INSURGENCY

The political and philosophical logic of the British Empire represented a reasonable basis for its existence in the mid-eighteenth century. Nevertheless, there was an aggressive cancer at work within the system that soon took over the logic of empire: economic gain. The mercantilist economy of Great Britain depended upon the influx of raw materials from its far-flung colonies and likewise needed the people on those distant shores to absorb England's manufactured goods. Each colony's performance within this system defined its value to London: productive colonies were carefully supervised and defended; unproductive colonies were largely ignored. New England was underestimated by imperial authorities who regarded them as competing with the same commodities produced in the British Isles. New Englanders had a large merchant marine—larger than any other colonial region—that also competed with the British Isles and engaged in smuggling. Accordingly, the New Englanders were subject to special attention post-1760 to regulate their shipping.[1]

To make the imperial system work, and to keep it independent of the economies of rival powers, the British Parliament enacted trade and navigation laws aimed at codifying the system. Colonies produced raw materials and bought manufactured goods. They were not permitted to engage in manufacture for themselves. Likewise, in England, citizens were to acquire the colonies' raw materials and convert them into goods for export. They were disallowed from producing, for example, tobacco or other raw goods, because to do so would disrupt the imperial system.

The American colonies, and New England in particular, played a key role in the system. Besides the production of raw materials, they also engaged in ship building, producing nearly a third of British shipping. They harvested the abundant fish of the North Atlantic, and they absorbed not only manufactures from England but also sugar and its byproducts from the English sugar islands in the Caribbean.

Complicating this relationship between mother country and her outliers was the issue of currency. In accordance with the prevailing economic theory of mercantilism, the British government sought to hoard specie within her own shores. This desire led to laws demanding hard coin from Americans in payment of debts, rather than allowing paper currency or notes. Coinage was very hard to come by, with the result that Americans were perpetually facing unmanageable debt. The only way they survived was through the "salutary neglect" of Robert Walpole's imperial authorities, who looked the other way as the enterprising American colonists traded enumerated commodities such as sugar and tobacco with non-British powers.

George Grenville brought the systemic flaws of empire to a head. He rose to the premiership in April 1763 and served in that role through July 1765. Uppermost in his priorities was the need to lead the empire in recovering financially from the enormous costs of the French and Indian War. Under his supervision, Parliament passed the Sugar and Currency Acts that gave rise to growing colonial resentment of the mother country. Because the resulting dissatisfaction did not express itself in a manner strong enough to thwart Parliament, the government followed with the Stamp Act in 1765. Whereas the former acts hit especially hard on merchants and the rich, the Stamp Act impacted everyone. It was interpreted as an illegal direct tax.

The Stamp Act was the spark that gave rise to the Patriot insurgency. Prior to 1765, even the most disgruntled colonial citizens took immense pride in living in the most prosperous, freest, and most powerful empire in the world—and an empire that had long left them largely alone. However, the Stamp Act brought to the surface the ultimately intolerable contradictions within the British Empire. Colonial gentlemen were theoretically free Englishmen who had never agreed to accept any lesser status. Nevertheless, Parliament viewed the colonies from the perspective of prevailing mercantilist theories—they existed for the benefit of the mother country. These two ideas could not coexist. The Patriot resistance arose to defend the former and neutralize the latter.

Pontiac's War

The destruction of New France and Britain's acquisition of Canada at the end of the war led to a radical realignment of politics among the Native American tribes of the *pays d'en haut* (upper country, consisting of the Great Lakes region), the Illinois country, and the Ohio Valley. Many of the Great Lakes and Illinois tribes (e.g., Ojibwe, Odawa, Potawatomi, Huron, Miami, Wea, Kickapoo, Mascouten, and Piankashaw) had enjoyed long-term trading and social relationships with the French. When the British assumed military and political control of Canada, they replaced the beneficent and respectful policies of past French Canadian governors with hubris and duplicity, alienating many tribes. Likewise, the Indians of the Ohio Valley (Delawares, Shawnee, Wyandot, and Mingo) had assumed that at the war's conclusion, the Europeans would stay clear of their land. Instead, the British strengthened their grip on their forts and trade routes.

The British commander in North America, General Jeffrey Amherst, believed that he could safely treat the Native Americans as a conquered people rather than as allies, because with the French now absent from the continent, Britain remained the sole dominant power. He ceased the expensive practice of ceremonial gift giving, which had previously underlain Indian diplomacy and patronage. Likewise, he and his officers did not disguise their cultural contempt for peoples they considered primitive and barbarian.

The Indians of the *pays d'en haut* rose up in rebellion against British authorities in response to the sudden change in policy and the interruption of expected gifts. Ottawa, Seneca, Mingo, and other tribes felt the affront the worst. Pontiac, an Ottowa chief, was only one of a group that led the attacks on British forts and outposts. The sudden violence was met with the arrival of more British troops, who eventually stabilized their position at some key forts (principally Fort Pitt and Fort Detroit). The British then entered negotiations with native chiefs that offered them sweeping concessions to restore the peace. The brief war underscored the deteriorating relations between the new British masters of Canada and the Native Americans.

Two key results followed Pontiac's Rebellion. First, the war was used as justification for continued British military presence in the colonies. The most immediate reason for the garrison was that Grenville was providing patronage for the suddenly unemployed military commanders after the French and Indian War. The redcoats also provided a handy way of coercing obedience from the increasingly rebellious colonists, but the ongoing threat of Indian violence was a convenient excuse for the soldiers' presence.

Second, the surprising outbreak among the Indians contributed to Parliament's decision to impose the Proclamation of 1763, which sought to forestall American colonists from settling Indian lands west of the Appalachians. The ministry wanted to avoid provoking the Indians any further, and they likewise were concerned that the further American settlers moved west, the less control the government would have over them. Unregulated migration westward might also spark European complications with the Spanish.

The Stamp Act Crisis

In March 1764, George Grenville proposed to Parliament that the ongoing budget crisis would require further revenue from the American colonies, and he put forth an idea—not yet solidified—that "certain Stamp Duties" would soon follow.[2] An uneasy correspondence ensued among representatives of Grenville's government, British agents in North America, and leading American colonists. On the one side was a desperation to solve the discouraging math of the British budget, coupled with constitutional hubris that brooked no compromise on the subject of Parliamentary authority. On the other was a stubborn resistance to government overreach that was in part mercenary and avaricious, and in part, founded upon the highest Enlightenment principles of liberty.

The response of one Jared Ingersoll, a citizen of Connecticut who had no sympathy for the Patriot radicals, shows how pervasive were the political ideals that eventually underlay the rebellion. He wrote to Thomas Whately, Secretary of the Treasury (and Grenville's main proponent for the Stamp Act legislation), that the American colonists "are filled with the most dreadfull apprehensions . . . [concerning]a tax laid upon a Country without the Consent of the Legislature of that Country and in the opinion of most of the people Contrary to the foundation principles of their natural and Constitutional rights and Liberties."[3]

Chapter 6. Patriot Insurgency

Figure 6-1: George Grenville, prime minister, 1763-1765.

Grenville's government and the Parliament in general had little sympathy for colonial misgivings about taxation. Debates in Parliament featured speeches about how the British government had "planted" and "protected" the American "children," and that the colonists should therefore be pleased to pay their fair share. Great Britain was at the height of its power, fresh from a stunning victory on a global scale, and its imperial hubris blinded the government from the perspectives of other Englishmen and the possibility of military defeat at the hands of their "children." The Stamp Act was duly passed in February 1765, signed by King George III in March, and delivered to the colonies in April.[4] It was a direct tax levied on legal documents, newspapers, and other items, including playing cards. The tax was scheduled to take effect on November 1, 1765.

How the colonies would respond to the Stamp Act related in part to the economic depression that followed the conclusion of the Seven Years' War. With the lucrative trade in foodstuffs and raw materials that the war had stimulated suddenly dried up, farmers, tradesmen, shipbuilders, and others found their livelihood threatened. The Currency Act of 1764 further squeezed the colonists, forcing them to pay their British creditors in specie, which was always in short supply. Patriot agitators pointed out that the Stamp Act would hit poor people especially hard because they would be forced to pay in coins that they did not have to bring forward legitimate lawsuits and other matters to the courts. Grenville acknowledged that the colonists might push back against the Stamp Act, but he anticipated no more than ineffectual protests, such as had occurred over the Currency and Sugar Acts. Indeed, he took steps to sweeten the bitter pill by requiring colonists to pay only two-thirds of what British citizens in England had to bear. Further, he directed that the monies collected would remain in the colonies to defray the expenses of the British forces there.[5]

Virginia was the first to respond formally to the offending act, led by a young upstart named Patrick Henry—the darling of his peers and an irritant for his conservative elders in the House of Burgesses. With the legislature partially dissolved for the season, Henry led the remaining members in adopting what became known as the Virginia Resolves. These were a series of six statements (though a total of eight were considered) declaring the principle that only Virginia's own legislature could levy a tax on Virginians, as a matter of British law and tradition. Henry's more inflammatory accusations of tyranny were voted down but reported in the press nevertheless. When the other colonies got wind of the Resolves, it goaded them into action, and soon similar measures were passed throughout the colonies.

In Boston, disapproval of the Stamp Act merged with a long-lived political feud between the supporters of the royal governor, Thomas Hutchinson, and his chief rivals, the Otis family. Hutchinson and his cohorts became associated with the Stamp Act and all that was odious about the British government. The Otis faction soon enlisted two infamous gangs within the city and set them on their course of mob violence. In August 1765, mobs destroyed numerous homes, burned and hanged effigies, and threatened the lives of Hutchinson and others connected to the Stamp Act duties collection.[6] The governor's home was thoroughly destroyed in an excess of violence that even the perpetrators admitted had gone too far; however, the point was made that the colonies would have none of the Stamp Act, nor of Parliament's presumptuous taxation.[7]

Beyond the general unwillingness to pay more taxes, some Patriot leaders noted with concern that if Parliament was permitted to directly tax the colonies, then the royal governors would have no incentive to call upon local legislatures for revenue. The Stamp Act, then, represented a direct assault upon representative government in the colonies—an institution that the colonists felt they had a right to as Englishmen. Although the revenue from the Stamp Tax was small, there were fears that once the precedent was set, the British would seek ever larger taxes. A larger revenue pool could then fully fund the colonial officials and render them independent from the financial control of the legislatures. The Patriot leaders' concerted response to Parliament's action pointed to bad news for the British, as the colonies appeared able and willing to work together, despite their sectional differences and primitive communications.

Boston's example served as a lesson to the other American colonies as to how to deal with the Stamp Act, its colonial sponsors, and the would-be tax collectors. Throughout North America, the crisis became a catalyst for open conflict among rival political factions. For those parties loyal to the Crown or who stood to gain from the Stamp Act, their association with Parliament's actions labeled them as enemies of the people. They quickly attracted insults in the newspapers and the targeted violence of the mobs. Effigies were hanged and burned, and collectors were threatened with vandalism or worse.[8,9] Within months of the Stamp Act's arrival in America, it had sparked widespread resistance. It remained to be seen if this spontaneous resistance against a piece of legislation could be unified and molded into a movement toward independence.

A gathering of delegates from nine of the colonies (Massachusetts, Connecticut, Rhode Island, New York, New Jersey, Pennsylvania, Delaware, Maryland, and South Carolina) in October 1765 seemed to answer that question. The so-called "Stamp Act Congress" met in New York City and issued declarations insisting that Parliament had no right to tax the colonies. The congress demonstrated that the colonial leaders were able and willing to unite under pressure and that—despite significant sectional differences—they shared the same philosophical resistance to Parliament's overreach.

Back in the mother country, the Stamp Act's sponsor, George Grenville, had fallen afoul of King George III over unrelated political matters. He had been dismissed as prime minister in July, and in his place, the Marquess of Rockingham headed up the new Whig government. Rockingham was besieged by American problems from the start. The depression of wartime trade coupled with the colonies' boycotts in response to the Sugar Act had already caused economic disruption among English merchants and creditors. With these new allies at his back, Rockingham moved to repeal the Stamp Act not on the grounds of constitutional principle, but rather as a means of rescuing the flagging economy at home. He made it clear to a Parliament deeply offended by American defiance that he fully intended to defend their sovereign rights over the colonies. However, the Stamp Act would have to go. In February 1766, the House of Commons voted to repeal it, but with this concession to American outrage came a legislative sister—the Declaratory Act.

The Townshend Acts

In the aftermath of the Stamp Act crisis, King George III dismissed Rockingham and asked William Pitt to form a government. To sweeten the deal, Pitt was made the Earl of Chatham, which effectively removed him from the action in the House of Commons and sent him to the Lords instead. Pitt put together an unwieldy administration including both friends and foes, but his own worsening health caused him to miss most of the key decision making. In his place, his Chancellor of the Exchequer, Charles Townshend, called the shots. Townshend was determined to put substance to the Declaratory Act by passing a new round of taxes on the American colonists. The resulting "Townshend Acts" levied new duties on incoming commodities, including paper, tea, glass, and other essentials. Parliament insisted that such duties were "external taxes" and belonged to its power to regulate trade within the empire. Because the colonists' main objections had been directed at "internal taxes," the new legislation was expected to proceed smoothly.

In addition to the new taxes, however, Townshend also moved to punish the New York legislature, which had rejected the Quartering Act requiring the colonies to provide housing and support for the British regulars stationed among them. The colonial objections to the act focused again on the issues of property, liberty, and consent. Quite apart from their aversion to standing troops, the New Yorkers protested that Parliament could not demand that they involuntarily give up their property to support the troops. To do so would be to once again attempt taxation without representation. This argument was entirely lost on the members of Parliament, who remained enraged at provincial challenges to their authority. According to the Suspending Act, until the legislators reaffirmed the government's Quartering Act, the New York assembly would be suspended and all of its acts nullified.

By the time the Americans responded to the Townshend Acts, their sponsor was dead. Charles Townshend died in September 1767, just as the colonial reaction to his legislation was picking up. John Dickinson's *Letters from a Farmer in Pennsylvania* insisted that the new taxes were a direct violation of the colonists' rights, just as the Stamp Act had been.[10] In response to these injuries, Dickinson recommended a quiet resolve to increase home manufactures, refrain from purchasing English goods, and the issuing of respectful petitions. Both he and the general public were exhausted from the violent upheavals that attended the Stamp Act crisis, and his call for peaceful, dignified resistance fit their mood.

The Boston legislature reacted to the Townshend duties with a "Circular Letter" to the other colonies urging unity and resolve to resist. It stated again in clear terms that Parliament could not tax populations not represented therein, and the colonies had no voice whatsoever there. The new taxes were intended not to regulate trade (which Parliament could do) but to generate revenue arbitrarily (which they could not do.) Through the remainder of 1767 and into 1768, the "Sons of Liberty"[11] in Boston opposed every effort by the royal governor to suppress their resistance. In desperation, he appealed to London to send in troops, and they did so, arriving in October 1768. Their presence provided a measure of security for the governor and other royal officials, but it also created a volatile atmosphere that might give way to violence that both sides hoped to avoid.

Meanwhile, the Circular Letter served as a catalyst for more political upheaval in the other colonies. The royal governors had received instructions to inform their legislatures that if the assemblies responded to the Circular Letter, the governors would dissolve them. The strong-arm tactics were ineffective. Informed by their earlier successes in the Stamp Act crisis, the colonial legislatures pushed back against royalist intimidation. As before, the struggle with Britain was framed by local political factionalism. The anti-Townshend Acts parties labeled their rivals as traitors and used the crisis to garner votes and more power in the assemblies. The royal governors in each colony continued in their strategy of trying to bully the legislatures by dissolving them. For their efforts, they earned yet another mass movement toward non-importation of British goods. Patriots all along the seaboard either officially or unofficially entered into agreements, eliminating imports of various goods and striving to stimulate domestic industries in their stead. Only Rhode Island held out against nonimportation. When they attempted to profit by distributing their British imports into the markets previously controlled by their colonial rivals, the other colonies began to shut off all trade with them. The Rhode Island merchants got the message and soon joined the nonimportation movement.[12]

The Boston Massacre[13]

Britain's decision to garrison the colonies was problematic on several fronts. Originally, the decision to do so after the successful conclusion of the Seven Years' War had more to do with providing patronage to British officers who needed a continued source of income.[14] The American colonists wondered why British regulars were needed now that the hated French had been expelled from Canada. There was some logic to the posting of troops along the western frontier, but it quickly became apparent that those troops were protecting the Indians from would-be white immigrants as much as they were protecting the colonies.

In the wake of the Stamp Act crisis, the ministry elected to station troops in the major cities along the seaboard, giving rise to the Quartering Act. Boston, New York, and Charles Town were each treated to the presence of redcoats. Popular antipathy rose against the garrisons because colonists saw the troops as an unwarranted expense as well as a threat to civil liberties. Englishmen on both sides of the Atlantic viewed the presence of a standing army through the lens of the English Civil War: it would inevitably be used for tyranny.

The modern US Army often engages in missions other than war—nation building, peacekeeping, disaster relief, etc. Experience has taught the necessity of keeping a firm hand on the behavior of troops in these scenarios. Combatant commanders therefore issue rules of engagement designed to restrain the soldiers' use of violence. Commanders also train their soldiers regarding the laws of war and peace. However, the British regulars in North America were deficient in their preparation for handling the surly colonists among whom they would be stationed. Patriots hurled abusive insults at soldiers, refused to give them off-duty employment and constantly tried to provoke them. Fistfights were common.

The soldiers' presence corresponded with reports and rumors of burglaries and rapes, further enraging the population. Even the well-behaved regulars became objects of wrath because many of them took advantage of the army's permission to seek civilian jobs when not on duty. Often, they accepted pay at a much lower rate than civilian laborers, thus depressing the labor market. When fights or other crimes were brought before the civilian courts in Boston, the military commanders discovered that the judges were prejudiced against them and even derided soldiers on trial, asking them what right they had to be in the city in the first place.[15]

Matters came to a head on March 5, 1770 near the customs house on King Street in Boston. It was a Monday following a week of more fistfights, threats, and insults on both sides. As evening fell, a redcoat named Private Hugh White struck a private citizen, Edward Gerrish, who had insulted the soldier's regiment. The lone soldier was soon surrounded by a growing mob. Captain Thomas Preston thereupon led a squad of six privates and a corporal to rescue White, but when the regulars arrived, they formed a line and leveled their muskets at the crowd. Preston tried to calm the mob, but someone threw an iceball, striking Private Hugh Montgomery, causing him to stumble. When he stood back up, he discharged his musket, and the other soldiers immediately began firing. Five Boston citizens died, and six more were wounded.

In the trials that followed that autumn, John Adams defended Captain Preston and did so successfully. Apart from two soldiers receiving light punishment for manslaughter, no others faced penalties for the affair. The incident played out in colonial newspapers as a rallying call against Britain's tyranny and the "illegal" use of troops in America. The British government could hardly have mishandled matters worse, and their haphazard use of military occupation helped push the Americans further into the camp of the Patriot insurgency.

The Tea Act, Boston Tea Party, and Intolerable Acts

Back in London, Lord North had taken over as prime minister, and his government moved to end the Townshend duties on everything except tea. North likewise modified the Currency Act in the colonies' favor, allowing them to use paper currency to defray public (but not private) debts. These developments seemed favorable to the colonies, and nonimportation soon ceased. Nevertheless, two issues that would prove fatal to reconciliation remained; the Declaratory Act was not overturned, and the tax on tea remained.

From 1770 through 1774, trade between Britain and the colonies thrived, but so did smuggling, customs inspections, cargo seizures, and the resentment of Parliament's tea tax. In 1772, a Royal Navy ship, the *Gaspee*, ran aground in Narragansett Bay while chasing a suspected smuggler. A crowd of Rhode Island citizens boarded the ship, injured the captain and crew, and then burned the vessel. The government responded with an investigating commission that proposed to take custody of any accused and send them to the mother country for trial. This violation of the right to trial by a jury of one's peers inflamed the colonies once again and led, in 1773, to the first serious consideration of independence in the press. The Virginia House of Burgesses thereupon decided to establish a Committee of Correspondance, with a role to establish ongoing communications with the other colonies. In short order, all of the other colonies except Pennsylvania followed suit. The Patriot network had been firmly established throughout the Atlantic seaboard.

The British government further enraged the citizens of Massachusetts and eventually the rest of the colonies by moving to insulate royal governors and judges from dependence on colonial legislatures for their salaries. The local assemblies felt that by providing pay for the royal officials, they could exercise a measure of control over them. Instead, under the new system, those officials would be paid from funds collected for the tea tax. This proposal brought to the fore all the old grievances against the Declaratory Act, the use of the military, and the long-held conspiracy theory that Britain intended to enforce Anglicanism on America. The Committees of Correspondence took up a feverish volume of letters, pamphlets, and newspaper articles protesting once again the allegedly tyrannical intentions of the mother country.

Thomas Hutchinson, now serving as Governor of Massachusetts, worsened the crisis through a series of letters he sent to Thomas Whately back in London. The letters, written from 1767 through 1769, expressed Hutchinson's strong support of Parliamentary sovereignty and his disgust with colonial rebellion. Most unfortunate for the future of British rule in America, he stated, "There must be an abridgment of what are called English liberties." He could hardly have done more to reinforce the conspiracy theories that framed all of the Patriots' misgiving about England. When Sam Adams and Thomas Cushing published the letters for all the public to see, a political storm broke. The Massachusetts assembly petitioned the government to remove Hutchinson.

At about the same time, Parliament passed the Tea Act of 1773. The British East India Company had operated since its founding in 1600 as a private corporation originally focused on trade in the Far East. By the time of the American Revolution, it had grown into an unofficial arm of the British Empire controlled only partially by the government. Its center of gravity became India, where in the course of the eighteenth century, it defeated its chief rival, France, and consolidated control of the eastern half of that region. The East India Company had its own private armies and administered the country by ruling over native power bases. By 1773, the company was strapped for cash, due to the increasing expenses involved in ruling India and the depression of trade that hit Europe after the Seven Years' War. In desperation, the company appealed to Parliament, many of whose members were stockholders, and the Commons duly passed the Tea Act. This legislation awarded the company a monopoly on the tea trade and retained the duties on it, all of which enraged the American colonists.

In combination with the Hutchinson Letters, the Tea Act revived the ardor and suspicions that the Patriots had toward the mother country. From their perspective, Great Britain was violating the constitution by acting in favor of the East India Company and at the expense of the colonies. When the tea-bearing ships began to arrive in American ports, they found that the merchants consigned to take the commodity had refused to unload it or pay the duties. Instead, they insisted—usually with the threat of violence—that the ships return to England with the tea. In New York, Philadelphia, and Charles Town, that is what happened. Unfortunately in Boston, the Patriots' plans to intimidate the merchants fell short. Governor Hutchinson would not cooperate and allow the three tea-bearing ships to depart, and the merchants were caught in the middle. In the end, they refused to cooperate with Sam Adams and the Patriots.

On the evening of December 16, 1773, the Patriots struck. Unidentified men dressed in Indian garb stormed the three ships, broke open the tea crates, and dumped the tea into the harbor. The "Boston Tea Party" destroyed about ten thousand pounds sterling worth of tea. This act of vandalism would lead directly to a crisis with England in a cascading sequence of events.[16]

When both official and unofficial news of the Tea Party reached London in January 1774, Parliament was outraged. The Privy Council summoned Benjamin Franklin and berated him for over an hour as he stood silent. Almost without exception, the incident left members of Parliament with the firm conviction that something had to be done to put the Americans in their place once and for all. Intricate matters of constitutionality and Enlightenment philosophy were brushed aside as the mother country determined to punish her wayward children. The legislative actions that Parliament undertook thereafter became known as the Intolerable Acts (or Coercive Acts) among the American colonists.

The first was the closure of the port of Boston until such time as the city compensated the East India Company for the destruction of the tea. The closure went into effect on June 15, 1774. The second was the Massachusetts Regulatory Act, which transferred some power from the local assembly to the royal governor, empowering him (instead of them) to appoint and dismiss government officials and juries. It also sought to limit town assembly meetings to one a year. The third act was the Impartial Administration of Justice Act that specified that any government officials accused of a capital crime would be removed either to England or to another colony for trial. Parliament also updated the hated Quartering Act, the new version calling for troops to be billeted with private families. Finally, the Quebec Act was legislation that favored the French settlers absorbed by conquest into British Canada. It provided that the Quebec Province could expand to include portions of modern-day southern Ontario, Illinois, Indiana, Michigan, Ohio, Wisconsin, and parts of Minnesota. It further protected French civil law and Roman Catholicism, even allowing churches to impose tithes.[17]

The colonial response to the Intolerable Acts reflected both the colonists' maturing political skills and the class and sectional differences that still threatened to divide them. Immediate calls for another round of nonimportation found merchants either vacillating or downright defiant. They argued that former nonimportation agreements had been violated by unscrupulous profiteers (John Hancock was one of the suspected culprits) and placed the lion's share of the burden on themselves. Still, many towns and cities took up the cry for boycotting English goods. At the same time, the colonies (except for Georgia, which was distracted by Indian troubles) sent delegates to the First Continental Congress in Philadelphia—demonstrating a powerful capability of uniting under pressure. To preserve that unity, the delegates agreed to give each represented colony one vote, regardless of size or population. Years later, in the 1787 Constitutional Convention, that balance would shift in a political compromise that accounted both for the power of states and for the weight of population.

A total of fifty-six delegates convened and found that they enjoyed a common understanding (nuances aside) of the constitutional issues at hand. Parliament, they concluded, did not have a right to tax them. At best, Parliament could exercise loose control over imperial trade issues. Both the government and the king were bound by law and tradition and had no calling to treat the colonists as inferior subjects. What to do about the Intolerable Acts was another matter. Trade restrictions were proposed, but so also was the idea of arming and training colonial militias in anticipation of armed conflict. In the end, the conservative voices prevailed, and the First Continental Congress issued a statement of its "Declarations and Resolves" as well as a "Petition to the King." Nonimportation was set to begin in December, to be followed by a ban on exports to England to commence in September 1775 if the offending acts of Parliament were not rescinded. A "Continental Association" elected in each town and city would enforce the measures. Most significantly, the delegates planned a second congress for the following year in the event that Great Britain refused their petitions.

Trade between the colonies and Britain dropped precipitously in 1775, but there would be no opportunity to determine if the Americans' economic sanctions would work because by the time the Second Continental Congress was scheduled to meet, war had already commenced. Governor and Commander-in-Chief Gage in Boston had urgently requested reinforcements to deal with what was becoming much more than just a discontented mob. The response from the government was that no reinforcements were coming but that Gage needed to meet force with force. In April, he made plans to do so.

Gage dispatched a force under Colonel Francis Smith to head to Concord, arrest Patriot leaders there, and confiscate stores of munitions. Early on the morning of April 19th, the scratch force of four hundred light troops and four hundred grenadiers departed Boston, intending to do so in secret. Fortuitously, local silversmith Paul Revere and a handful of other Patriots noted the early morning movements and set out to rouse the countryside. Soon militia companies and "minutemen" were turning out of towns, villages, and farms. An initial skirmish at Lexington between Major John Pitcairn's light troops and Patriot militia Captain John Parker did not auger well for the rebel cause. The British regulars opened fire and managed to kill eight and wound ten more as the Patriot militia melted way after hardly firing a shot. Thus encouraged, the British continued to Concord and ransacked the town looking for weapons. They inadvertently set fire to some buildings, and this act seemed to goad the gathering Patriot militias into action. For the rest of the day, they followed the retreating redcoats back to Boston, taking every opportunity to ambush and snipe at them. The Patriots might well have inflicted a decisive defeat on Pitcairn's troops were it not for the arrival in Lexington of Brigadier General Earl Percy with a thousand reinforcements. The united British force rested briefly and then headed back for the safety of Boston, but the Patriots ambushed them all along the way. When the exhausted regulars reached the city, they had suffered 273 casualties and inflicted but ninety-five on the Patriots.

News of the battles at Lexington and Concord spread quickly throughout the colonies, and soon the nearest colonies were sending militia companies that camped outside of Boston, cutting the troops within off from the interior. In May 1775, the Second Continental Congress met, but despite the bloodshed, the delegates represented a wide spectrum of opinion regarding what should be done. They were not yet ready to make the decisive break with England, but they were prepared to authorize an army, and they did so on June 14th. George Washington was given command, but before he arrived, the half-organized Patriot army gathered around Boston had been blooded at the Battle of Bunker Hill. Technically a British victory, it had come at the cost of 226 regulars killed and another 828 wounded.

The Second Continental Congress took on the role of a national government—a role generally accepted among the thirteen colonies. They began the difficult business of administering the war effort and seeking negotiations with Great Britain in the hopes that a final break might be avoided. They issued the "Olive Branch Petition" as an offer to the king to negotiate, but by the time the document reached England, it was too late. Passions won out on both sides, and the violence continued, leading belatedly to the question of independence. The king refused to read the petition and instead declared the colonies to be in rebellion, and he labeled the perpetrators as traitors. The Congress also issued the "Declaration of Causes and Necessity of Taking Up Arms," which enumerated all the constitutional arguments that had been developed since the Sugar and Currency Acts of 1764.

The impulse toward independence drew strength from two primary factors. The first was the collective experience of the last decade of political conflict that made clear that neither king nor Parliament had any intention of making constitutional concessions to the colonists. The second was the urgent need to achieve a military alliance with a foreign power that would join in a war against Great Britain. As long as the colonies remained possessions of Britain, they could not conclude treaties. Hence, the Declaration of Independence, when it came, was in effect the necessary prelude to international diplomacy. Serious discussion of the matter began in May 1776, and on July 4, the Second Continental Congress approved the text of the document that formalized the break with England. The Patriot insurgency had succeeded in declaring the thirteen colonies a separate country.

LEADERSHIP, ORGANIZATIONAL STRUCTURE, AND COMMAND AND CONTROL

The Patriot insurgency began as a political movement and grew from the time of the Stamp Act crisis of 1764 into a network of political resistance against what was perceived to be British tyranny. Patriot leaders came from the political elites of the thirteen colonies, especially Massachusetts, Virginia, Pennsylvania, and New York. They did not organize themselves in a pattern matching modern insurgencies, primarily because in the colonies, governmental structures already existed. Thus, colonial legislatures deputized key individuals to act as delegates to the Continental Congress, and the Congress in turn relied on these men to administer the government, conduct diplomacy abroad, and command the embryonic military forces.

Underground Component and Auxiliary Component

In modern American military doctrine, the underground is that part of an insurgent movement that operates in places where the armed component cannot operate. Typically, an insurgent underground is found in urban areas. As the Patriot insurgency developed, the cities and towns of New England and Virginia—and eventually all the colonies—became the loci of political and financial resistance, while the countryside played the important role of sustaining the rebellion. The organizational structure most suited to frame that resistance was the colonial legislatures and municipal assemblies. When the royal governors dissolved the assemblies and formally disallowed them to meet, the colonists responded with "Committees of Safety" and similar organizations that conducted activities known today as "shadow government"—governing the people illegally and ignoring the official British government. At the same time, the colonial undergrounds established "Committees of Correspondence" as a network of insurgents that sought to bind the thirteen colonies together in purpose. By 1777, the Patriots had ousted royal authority across the colonies. All royal governors (except in New Jersey) were run out of office, and Loyalist members of colonial assemblies and members of the press were silenced. Patriots had control of nearly all militias, and the remaining Loyalists in Virginia, the Carolinas, and Georgia were either defeated in battle or forced to take oaths of neutrality. Hence, the Patriots' shadow government was the only remaining authority in much of colonial North America.

The role of auxiliaries was left to the people inhabiting the thirteen colonies. Histories suggest that, in rough numbers, about one third embraced the Patriot ideology and agreed with the independence movement, one third opposed and wanted to remain loyal to Britain, and one third were undecided. However, in the course of 1776, two documents effectively sealed the domination of the Patriots: Thomas Paine's *Common Sense* and the Declaration of Independence. By the end of the year, although prospects for a victorious outcome were much in doubt, there was a greater feeling of unity of purpose in breaking with England. The colonists supported the Patriot cause when asked or intimidated into compliance, and not a few Patriots voluntarily gave very generously to support Congress and the Continental Army.

Armed Component

One of the most distinguishing features of the Patriot insurgency was the leaders' choice of how to build an armed component. In a typical modern insurgency, leaders most often opt to build a guerilla army composed of cells and small units that gradually coalesce into ever more powerful irregular forces. If they are successful, those forces may gradually morph into conventional forces toward the end of the insurgents' struggle.

In the Patriot insurgency, however, leaders opted to immediately form conventional forces (along with some small but significant irregular forces). Those conventional militias—and later the Continental Army—were intended to fight on conventional battlefields using conventional tactics, in the hope that they could stand up to British regulars and Hessian mercenaries.

Facing off against trained, disciplined, and well-equipped regulars with militia regiments that often lacked equipment, training, and ammunition was a risky enterprise, which even the Patriot leaders at the time recognized. Why then would they choose to form conventional units instead of resorting to guerilla armies? To understand why they chose to develop the armed component they did, it is important to comprehend the nature of the resistance movement, the geography of the theater of war, and the experience and education of the officers and legislators (especially Washington).

The Nature of the Resistance

The Patriot insurgency developed politically from within established colonial governments. The town assemblies and colonial legislatures operated from within an established infrastructure of buildings, laws, and economies. They did not begin their efforts while hiding out in mountain refuges, but rather while debating in courtrooms and congresses. The men who formed the leadership were mostly landowners and merchants with valuable estates that needed armed protection. Thus, the men who called up and led the insurgent armies were predisposed to conventional approaches to combat. As colonial gentlemen, they possessed a natural dislike for what they would have considered rag-tag, illegal armed groups. In short, established infrastructure requires conventional forces to defend it.

Second, the Patriot leaders did not start from scratch in building the rebel military establishment. Long before the conflict began, colonies had formed militias. Some were hardly more than social clubs, while others—the Virginia Regiment, for example—more closely resembled the organizational structure of a regular regiment. Traditions, enlistment rolls, and a proud (if mixed) military history were already on hand. Washington and his officers built new regiments for the Continental Army as well, but the norms of military organization pointed them toward conventional units rather than guerilla forces.

Finally, the Patriots had to convince their countrymen of the strength and respectability of their cause. The Continental Army was an instrument of public diplomacy as well as a military tool. It comported with what would be expected of an aspiring nation. An army of only irregulars would be susceptible to being labeled criminals and ruffians. Instead, Washington and his officers hoped to build an army that was both effective and respected.

Geography of the Theater of War

The geography of the American colonies helped to define the character of the Patriot armed component. To put it simply, to the west was a frontier featuring rough, wooded terrain, few roads, and lots of Indians. To the east, across the Atlantic, was Europe—with the most advanced infrastructure on the planet and the most highly trained conventional forces anywhere. In between was the Patriot insurgency that would have to dominate—or at least survive—both challenges.

The American colonies could not defend their frontiers (or satisfy the ambitions of the land-hungry speculators) without irregular forces that could fight the Indians over rough terrain. Washington and his Virginia Regiment were exemplary frontier fighters, used with varying degrees of success by British generals in the French and Indian War—chiefly Edward Braddock and John Forbes. The American irregular forces knew how to navigate, build roads, scout for Indians, and fight through ambushes when necessary. Thus, from an early date, American military tradition included irregular warfare experience.

Nevertheless, to fight the British, those same colonies would have to be able to form conventional armies that could face off against disciplined ranks of redcoats and match them musket for musket. Whereas the frontier rifle was as effective as the marksman employing it, smoothbore muskets required a high volume of fire from close-packed ranks of trained infantry. In the pre-Revolution generations, colonial militias trained to supplement British regiments, and colonial gentlemen longed for a coveted commission in the King's Army. Consequently, American warfighting traditions merged the frontier ethos with conventional military theory and practice.

Experience and Education of Patriot Officers

George Washington had served as an officer in the Virginia militia in support of regular British forces. He nursed an ambition to become an officer in the British Army, and he naturally thought about military affairs from within that paradigm. Examination of his writings points to an evolving attitude toward irregular forces. During his command of the Virginia Regiment in the French and Indian War, Washington boasted of his troops' frontier savvy. Throughout his career as commanding general of the Continental Army, he often expressed his contempt for irregular forces and volunteer militias as unreliable and poorly disciplined.

Thus, from the start of the resistance movement, it was inevitable that the insurgent leaders would opt for a conventional approach to war. As the conflict ensued and lengthened, the exigencies of combat required them to integrate irregular forces into the overall military structure, but they did so reluctantly and, given the chance, always reverted to conventional forces and conventional tactics.

Public Component

The Patriot insurgency employed two major forms of public communications. The most direct communications emanated from state legislative assemblies and the Continental Congress. The assembled leaders communicated, often through their network of Committees of Correspondence, by issuing resolutions, agreements, and non-binding calls for nonimportation, boycotts, and other measures of resistance.

Second, the Patriots got their message out through newspapers, journals, pamphlets, and tracts. Journalists on all sides of the rebellion issue were active throughout the conflict. The most notorious and effective among them—both Patriots and Loyalists—sometimes attracted ire and violence against them, but they were effective in providing a public forum for debating the weighty issue of whether to break with the mother country. Early in the conflict, the Patriots gained control over the press and suppressed Loyalist newspapers.

Newspapers

Among the most influential newspapers prior to and during the American Revolution was the *Boston Gazette*, published by Benjamin

Edes and John Gill starting in 1755. The port of Boston had immediate access to all the latest news from England and the rest of Europe, and the *Gazette* served as a primary source of that information for most of New England. It also framed the interpretation and discussion of key events, including the Boston Massacre, the Boston Tea Party, and the city's response to the Coercive Acts. Samuel Adams used the *Boston Gazette* to urge resistance to British measures, and the newspaper helped propel and "normalize" Adams' rhetoric. The paper's chief rival was the *Boston Evening Post*, which often served as a vehicle for expressing the royalist position on political matters. Likewise, the *Boston Chronicle*, printed by John Mein, carried articles hostile to the Patriot cause.[18]

The *Pennsylvania Journal*, published by William and Thomas Bradford in Philadelphia, was likewise an important voice for the Patriot insurgency—silenced only when British regulars were occupying the city. In 1765, the Bradfords changed the masthead of the journal to resemble a gravestone to symbolize Great Britain's efforts to kill American liberties.

Figure 6-2: The *Pennsylvania Journal*

The *Massachusetts Spy*, published by Isaiah Thomas, relocated from Boston to Worcester in 1775 and became famous as the premier account of the Battles of Lexington and Concord. Thomas used the occasion to emphasize that "British troops, uumolested and unprovoked wantonly, and in a most inhuman manner fired upon and killed a number of our countrymen."[19] Throughout the revolution, Thomas remained a strong voice for the Patriots.

From 1775 through 1782, Benjamin Towne published the *Pennsylvania Evening Post* in Philadelphia, even during British occupation. The newspaper published important articles from the Continental Congress, including the views of individual members. In 1776, it led the way for the publication of the Declaration of Independence, printing the document in its entirety. Other papers throughout the colonies followed suit. Because Towne was able to maintain the appearance of evenhandedness and was diplomatic in his handling of both Tories and Patriots, he was able to remain in operation throughout the conflict.

The *Connecticut Courant*, published from 1769, became an influential voice for the Patriots because it continued to operate from Hartford when other cities were occupied by the British (which had the effect of shutting down Patriot newspapers). With the British in New York City and Philadelphia, the *Courant* was the newspaper of choice for many in the Middle Colonies and beyond.

From 1776 through 1784, Patriot John Holt published the *New York Journal*. He began operations in New York City but fled from there to successive locations to stay clear of British authorities. His newspaper was a strong voice for independence and was valued for its insights concerning the military and political situation in New England.

The *Providence Gazette*, published in Rhode Island by William and Sarah Goddard and later owned by John Carter, championed colonial rights and independence from the early 1760s. The publishers insisted that history was moving toward American independence and unabashedly urged its readers to participate in the inevitable break with England. Like other journals, its circulation benefited from the British occupation of Boston and New York as those cities' Patriot newspapers were forced to shut down.

IDEOLOGY

Religious Influences

The Patriots' ideology was framed by their (for the most part) common Protestant religious background. The various denominations had roots in the Protestant Reformation and the English Civil War. Consequently, the Christian worldview that the Patriot leaders espoused drew heavily from Anglican, Calvinist, and postmillennial roots. The early Puritan settlers in New England established the precedent of Congregational church government, as opposed to an episcopal or Presbyterian system.[20] This system of church rule by the congregation members themselves helped to establish the ideas of limits on human authorities and the importance of consent and representative government. Congregationalists were instrumental in the founding of both Yale and Harvard universities, and they were also prominent influences in the First and Second Great Awakenings.

Postmillennialism refers to the eschatological belief that God intended to use Christians—as the Body of Christ—to establish the Kingdom of God on the earth. Christians would lead the world through virtue and the gospel of Christ toward the establishment of a just and righteous government that would reflect the ideal attributes of New Testament Christianity. This process would be gradual and historical (i.e., not apocalyptic), and some of the Founding Fathers viewed the American Revolution through that lens. John Adams, in particular, believed that the Patriots were completing the unfinished work of the Reformation and building the ideal Kingdom of God in North America. It was to be a "kingdom" only in metaphor. In actuality, it was to be a republic of virtuous Christian men, and the resulting government would serve as a beacon of freedom to the world.[21] John Adams held complex views on the matter. He publically respected New England's religious heritage but privately was closer to a deist.

Indeed, despite that fact that Jesus Christ instructed his disciples to pray "Thy Kingdom come, Thy will be done on earth as it is in Heaven," some Patriots viewed monarchy as a manifestation of heathenism, until support from the King of France was later offered. Thomas Paine's *Common Sense* claimed that the whole idea of kings ruling subjects emanated from paganism and that as true Christians, Americans must throw off the British monarchy in accordance with the "laws of nature and nature's God."[22] The common religious framework shared by nearly all English colonists in America led them to characterize their political enemies as "evil" and themselves as "good." Their natural assumption, then, was that God was on their side.

The democratic undertones of Protestantism in general and Congregationalism in particular merged well with Enlightenment principles.

This confluence of philosophy and religion was perhaps best illustrated in the person of Jonathan Edwards (1703-1758), an influential theologian and writer in pre-Revolutionary America. He was instrumental in stimulating the "Great Awakening" of 1735-1744. His most famous sermon, "Sinners in the Hands of an Angry God," led to many conversions and greater attention to moral behavior as fundamental to the success of a free society. Edwards' impact on the revolution grew in the next generation from the legacy of his writings. In his own time, George Whitefield was a more important contemporary.

The Great Awakening was a widespread religious revival that had its roots in England but grew into a major societal influence in America. George Whitefield, an evangelical Anglican preacher (of the sort that later became Methodists), came to the colonies in 1740 and delivered a series of sermons calling for spiritual revival. The staid, unemotional piety of the Congregational churches he visited was soon supplanted by shrieking, enthusiastic believers anxious to experience the fullness of salvation. Within a few years, the movement began to threaten established authorities within the churches themselves, resulting in a bifurcation of opinion regarding the new revivalism. "Old Lights" (i.e., the established church authorities) rejected the religious innovations while "New Lights" embraced them. This factional dispute transcended mere religious dogma and bled over into daily politics. As the revolution approached, most Congregational and Presbyterian ministers sided with the Patriots and against what they perceived as evil British (and Anglican/Catholic) tyranny.[23] Even so, most Anglicans supported the Patriots.

Good King, Bad King

The evolution of American sentiments toward King George III serves as a reliable indicator of how and when the ideology of independence began to dominate affairs in the colonies. Until 1776, even the most ardent Patriots thought of the king as their ally against the predatory Parliament and royal governors. The good King George, they believed, would ultimately come to the rescue of the colonies and restore justice. It was only in 1775, when that very king went out of his way to demonstrate his animosity toward the Patriot cause that the Americans came to realize that George III was part of the problem.

King George III ascended the throne of Great Britain in 1760. His accession was celebrated in the American colonies because the king represented—in the minds of the colonists—all that was good about the empire. Protection from European enemies, easy connection to the global economy, the continued domination of white Protestants over everyone else. The seventeenth century "Glorious Revolution" in England ensured that the sovereign would remain not only Protestant but also limited in his powers. The resulting "King-in-Parliament" arrangement served to check excesses that might damage liberty. Further limitations on royal power came in the form of colonial assemblies elected by landowners. These mini-legislatures were granted the power to tax the colonies, and the consequent balance of power between London and locals served to keep relations between mother country and colonies relatively peaceful.

That all began to change with the Stamp Act crisis, but in the resulting political conflict, American colonists continued to view King George as their advocate against a rapacious Parliament. Yet in the ten years that separated the Stamp Act Crisis from the Declaration of Independence, the Patriot position regarding their sovereign changed radically. Delegates of the first Continental Congress (1774) described the king in less glowing terms than before and portrayed him as an unwitting victim of his evil councilors. The colonists hoped that George III would stir himself, oust the bad actors advising him, and bring Parliament to heel. Instead, in the crisis following the Intolerable Acts, the king showed himself their vigorous defender. Consequently, the text of Jefferson's Declaration two years later focused not on Parliament, but on the king himself: "The history of the present King of Great Britain is a history of repeated injuries and usurpations, all having in direct object the establishment of an absolute Tyranny over these States." This switch of antipathy from the Commons and Lords to the person of King George III signaled the death of Patriot hopes that revolution could be avoided and that the king might right the ship of state without further violence. Once the inviolable dignity of the king was swept away, the path to radicalism and revolution was open.

The Meaning of Liberty

It is easy to look back on the late eighteenth century and imagine that the Patriot leaders were fighting for "liberty" as we understand the term today. However, liberty was a very different concept to the Founding Fathers than it is today. Most fundamentally, the word "liberty" had everything to do with property. A person without property or—even worse—a person whose property could be arbitrarily seized by the government could not be said to have liberty. Instead, liberty meant the right to own property and dispose of it as one sees fit without interference from the government. Patriot leaders drew from Enlightenment philosophy to sharpen their understanding of the concept. John Locke's *Two Treatises of Government* in particular set down the most popular explanation of "social contract theory" that championed consent as the basis for legitimate government. According to this expression of Whig philosophy, free-born men of property willingly consented to submit to the government, which in turn was obliged to protect the liberty of the consenting. Regrettably, "property" in the eighteenth century included slaves. Hence, Patriot ideology concerning liberty implied protection of slave owning as well.

The Sugar Act and the Stamp Act that followed represented assaults on property and therefore on liberty. The English constitution, after all, was built on the ideal that taxes could not be extracted but rather had to be granted by a representative body. Because Americans were not represented in Parliament, they could not therein agree to a tax. In the words of William Pitt: "We your Majesty's Commons of Great Britain, give and grant to your Majesty, what? Our own Property? No. We give and grant to your Majesty, the property of your Majesty's commons of America. It is an absurdity in terms."[24] Parliament's argument that the American colonists were "virtually represented" by members whom they did not elect fell on deaf ears in North America.

To garner sufficient combat power to win the war, Patriot aristocrats had to reach down into the middling and lower classes, and to do so, they conceded a new form of "liberty" that would come to mean economic opportunity for all and civil rights that would apply to all free men—not, however, to women or the enslaved.[25] Thomas Paine published *Common Sense* in 1776—a pamphlet aimed at the common people that sought to demonstrate that the British government was unnecessary and destructive in North America. Paine facilitated the evolution of the concept of liberty from a notion related to property ownership to a more general idea of civil and human rights—at least for white, Protestant males.

Conspiracy Theory

Revolutionary ideology grew in part from a widespread belief among the colonists that king and Parliament were operating as part of a grand conspiracy against the interests of the innocent, hard-working, godly citizens of North America. The vitality and reach of the various newspapers and journals throughout the Atlantic seaboard propelled conspiracy theories toward everyone who could read. Whereas colonial legislatures and officials had to maintain a restrained and dignified composure, journalists were free to give rein to passion and accusation. The result was a growing culture of paranoia and suspicion regarding the intent of the British government.

John Stuart, Third Earl of Bute (1713-1792), lay at the heart of American conspiracy theories. A Scottish nobleman, he became young George III's principal tutor and the most formative influence on the future king. He schooled his protégé on the importance of becoming, in the words of Henry Bolingbroke "the Patriot King"—a monarch who served the true interests of the empire, isolated from the petty politics of Parliament.[26] In practice, this perspective demonstrated resistance against the evolving English constitution of King-in-Parliament and suggested that when George III ascended the throne in 1760, he might assert monarchical power against the interests of the Commons. Bute stimulated powerful resistance among the Whigs in Parliament, and even after he resigned from public life in 1763 after a brief but stormy turn as prime minister, there was a pervasive suspicion that he was pulling the strings behind the throne. As a consequence of his unpopular policies (including the infamous Cider Tax his government imposed on England), he became the most hated politician of his day. The Patriot insurgents attributed nearly every evil of the British government to his nefarious influence. Local Tories targeted by Patriot mobs were often hanged or burned in effigy with a boot tied around their necks—a homonymic reference to the earl. Bute was also associated with Jacobitism—the political movement that tried to have James II or his descendants restored to the throne as a Catholic monarch. Hence, conspiracy theories tied to Bute typically asserted that the British government was working to suppress Protestantism and force Roman Catholicism (or at least a Catholic version of Anglicanism) on the colonies.

Conspiracy theory was a powerful motivator toward revolution among the common people. Newspapers, pamphlets, journals, and letters provided the Americans not only their news, but also their entertainment. By purporting various shadowy conspiracies, they facilitated a catharsis of hatred and frustration toward authority in general and Parliament in particular. The seemingly far-fetched notions that Bute and his co-conspirators were attempting to enslave Englishmen at home and in America found their way even into such vaunted Patriot documents as the Declaration of Independence, which stated that the record of the king's actions "is a history of repeated injuries and usurpations, all having in direct object the establishment of an absolute Tyranny over these States." Conspiracy theory allowed even educated aristocratic Patriots to ignore the economic and political factors that underlay the government's actions and instead attribute them to the evil purposes of a few powerful men.

This Land is My Land

Another key aspect of Patriot ideology had to do with the abundant lands in North America. Because liberty in the minds of political elites had everything to do with land ownership, the colonial upper class and middle class craved land for themselves and their many children. They lived on a continent that offered opportunities to acquire such land, limited only by the minor annoyance of having to remove the Native Americans who owned it.

When the Proclamation of 1763 slapped what appeared as an arbitrary and unjust limitation on future land acquisition, it cut into several layers of colonial society. Poor farmers took offense at a presumptuous, distant government trying to restrict their access to western land. Middling sorts saw their hopes of buying or leasing productive farmlands go up in smoke. Land speculators were aghast at the prospect of losing substantial revenue, and the land-owning elites were outraged at such a brash assault on liberty. Much of the resulting rebellion sought to widen access to land ownership after the king tried to restrict it.

Choose a Side

As the Stamp Act and Townshend Duties crises played out, a clear pattern emerged that helped to sharpen the evolving Patriot ideology: the need for citizens to choose a side. Prior to the war, the Patriots' chief means of resistance were economic—boycotts and nonimportation—but these methods relied upon widespread participation for effect. If one group of merchants or consumers failed to follow the economic sanctions, then British merchants would not feel the squeeze nor be able to influence Parliament to undo the offending legislation. Consequently, Patriots resorted to measures designed to encourage or, if necessary, compel compliance. Committees of Inspection and similar extra-legal groups took it upon themselves to investigate merchants suspected of importing British goods. The threat of mob violence reinforced the point for anyone wavering in their patriotic duty.

The nature of local politics led to rival factions branding each other as pro-British or anti-British, ignoring the underlying subtleties. Thomas Hutchinson in Boston, for example, was actually against the Stamp Act, but because the rival Otis faction saw an opportunity for political gain, Hutchinson was labeled as a royalist and abused as an unpatriotic, conniving, dishonest, and devilish conspirator. Extreme condemnations on both sides served to sharpen the major dividing line between them while blurring the many other divisions along social, religious, and economic lines. As the crucial year of 1776 approached, leaders of both the Loyalists and the Patriots pushed to simplify the options for every citizen: you are either with us or with them. The publication of the Declaration of Independence clarified the bifurcation even further. It was increasingly difficult and dangerous to be neutral.

The Question of Independence

As late as 1775, Benjamin Franklin famously observed that he had never heard anyone in America expressing a desire to separate from Great Britain. His statement was hyperbole, but it served to illustrate that despite a widely felt dissatisfaction with British policy toward the colonies, the people were disinclined to rebellion. In the first place, many American colonists benefited from the British economic system. The Royal Navy protected a global market that offered outlets for produce and raw materials and the luxury of imported finished goods. Second, Great Britain protected the colonists from other European powers—especially France—who might otherwise invade and conquer them. This protection became much less valued after the forcible removal of the French after 1763, but it remained a factor in the mind of the average English colonist. Finally, American colonists understood that separation and independence would mean war against the most powerful nation on earth. It was hard to imagine how thirteen separate, somewhat primitive colonies would be able to generate enough combat power to prevail in such a contest. To lose would mean harsh reprisals from London.

The Patriots, therefore, had their work cut out for them. They had to appeal to the populace in such a manner that they could emphasize the *rightness* of the cause as well as the *feasibility* of the plan. It was easier to make fellow citizens feel offended than it was to conjure a popular vision of a colonial strategy that would create and defend a new nation.

MOTIVATION AND BEHAVIOR

The Patriot insurgency faced a complex and dangerous set of challenges as their leaders slowly contemplated the question of independence. To make such a daring break with the mother country would involve the Patriots in not one but three wars, and potentially a fourth.

First, they would have to survive the expected onslaught from Great Britain. The Royal Navy would dominate the sea lanes and facilitate the British seizure of America's major ports, but from there, the problems facing king and Parliament would become far more intractable, for the real question would be: could the British, through military operations and political coercion, destroy, capture, or otherwise neutralize the Patriot insurgents? Thus, the Patriots' first challenge would be war against Mother England.

Second, the Patriots would have to deal with the Indians. In a classic reworking of "the enemy of my enemy is my friend" logic, the Royalists found themselves in an alliance of convenience with the Indians of the Ohio Valley, the remnants of the Iroquois Confederation, and the *pays d'en haute*. The various tribes were desirous of keeping the American colonists east of the Appalachian Mountains, and as it happened, so were the British. The Proclamation of 1763 made an alliance between the British and the Indians a genuine coalescence of interests: to keep the Americans from expanding to the west, which would divest the Indians of their lands and encroach upon the interests of British Canada (which was populated largely by French Catholics). The Indians lacked trade goods and weapons that the British could supply; the British lacked the skilled guerrilla warriors in the interior that the

Indians could provide. The Patriots, then, would have to defeat the Indian tribes as well as the British redcoats.

Third, the American Revolution—if there was to be one at all—would necessarily be a civil war. Loyalists, also called Tories, comprised at least a fifth of the colonial population. Loyalist leaders, landowners, slave owners, and pensioners were heavily invested—similar to the property portfolios of the Patriots—in the continuance of British rule. To see the colonies separated from the mother country would mean not only political chaos but also financial ruin. As the civil war played out and became ever more bloody, American independence might also mean death or imprisonment. The Loyalists took on the role of counterinsurgents, whose job was to resist the Patriot milias.

Finally, there was the potential fourth conflict: class warfare. As described earlier, the very concept of "liberty" held different connotations to different groups of people, including those who gathered under the Patriot banner. Which vision of freedom would prevail? They could not coexist because the elites' version of liberty included the perpetual subordination of lower classes to the aristocracy. The "middling classes'" interpretation of liberty included liberation from the artificial bonds of deference to gentlemen ranging from King George to George Washington. Among the most delicate problems facing the Patriot leaders would be their management of the socioeconomic implications of independence. In the end, the leaders needed support from the lower classes—artisans, farmers, and urban poor—to serve in their armies, and they would not be denied those tenets for which they fought. The resultant Constitution of 1787 would represent, among others, a compromise between the elites and the people—a compromise that avoided a bloody class war.

The Character of the Revolution

Much of the popular history of the American Revolution, written as the United States grew to be a world power and then a superpower, characterized the American Revolution in sympathetic and admiring terms. When compared with the French or Russian revolutions, the American version was seen as relatively peaceful, respectable, and orderly. In reality, it was exceedingly bloody. Americans killed Americans to a degree matched only in the Civil War nearly a century later.

Patriots often coerced support for revolution by publicly shaming Loyalists, labeling them as traitors—a characterization that would excuse violence against them. As a consequence of the war, sixty thousand Loyalists became refugees. One fifth of the population remained enslaved during the great struggle for liberty—an irony that bedeviled the Patriot leadership to their graves. The true character of the revolution could only be accurately described as profligate violence—if not quite on the scale of the European revolutions.[27]

Elites and Commoners

Early historians of the American Revolution fixated on the role of the political and social elites whose names feature prominently in textbooks: George Washington, Thomas Jefferson, John Adams, Benjamin Franklin, and others. Analysts within this historiographical tradition focused on intercolonial organizations—primarily the Committees of Correspondence and the Continental Congress. The political, military, and ideological leadership of the elites was seen as the decisive catalyst that led to successful revolution. From this viewpoint, one might conclude that the only Patriots of consequence were those whose faces appear in museum portraits.

The past half century, however, featured the growth of alternative historiographies, including a focus on the common people of a given era. The American Revolution has attracted the attention of historians in this tradition, and they tend to challenge the monopoly of elite leadership in the movement. Instead, this perspective examines the general population and concludes that the commoners who engaged in political agitation, participated in boycotts, and eventually manned the Patriot regiments were motivated by a combination of factors that had little to do with Jefferson's high-sounding words or Washington's dignified leadership. Evangelical religion combined with a general sense of natural rights led farmers, merchants, dock workers, and others to resent an unelected imperial government that seemed determined to harm rather than protect the people. Scholars of the social history viewpoint downplay the role of urban elites and look to rural commoners as the center of gravity of the rebellion.[28]

An objective integration of these two approaches to historiography yields the truth: both elites and commoners played a role in forming the Patriot insurgency. Educated, aware, and passionate commoners sustained and manned the insurgency. The power of their combined closed purses shook the British economy. They paid the taxes that funded the armies and congresses. They shed the blood that led to ultimate victory. However, the elites also played a crucial role. The insurgency would have accomplished little or nothing without their leadership. It was the elites who communicated with the king and Parliament, and it was also the elites who formed and framed the intercolonial cooperation that eventually prevailed. The winning combination for the Patriot insurgency was widespread willingness to risk all for liberty pulled together and led by an educated, effective team of political elites. One could not do without the other, and together they generated enough political energy to endure and outlast British aggression.

OPERATIONS

Paramilitary

Land Operations

The Patriot military strategy was centered on land operations. The insurgents' objectives included the elimination of Loyalist influence, attrition of British forces, preservation of the Continental Army, protection of colonial cities and countryside, and eventual military control of the entire thirteen colonies. To achieve these objectives, Washington and his subordinate leaders had to cooperate with the French armies and navy and resist Great Britain's Indian allies. A thorough military history of the revolution exceeds the scope and intent of this work. Instead, this section briefly reviews the course of land operations that led to the eventual capitulation of British forces in America.

Viewed as a whole, the land campaigns of the American Revolution were not in themselves decisive. At best, the Patriots could hope to eliminate Loyalist strongholds and push back Britain's Indian allies. They could also induce the French to join the war effort by demonstrating that the British armies would not easily sweep away the rebels. However, they could not, through ground operations, inflict a crippling and decisive blow to Great Britain. Instead, the land campaigns gradually wore down the British forces, increasing London's costs as the war prolonged. The war in North America likewise acted as a political focal point: the continued existence of the Continental Army gave the Patriots a palpable demonstration that the revolution continued. The need to support and direct the army gave the Continental Congress a potentially unifying problem to solve. However, to win the conflict for independence, the Patriots would have to force Parliament to give it up. Because Parliament was largely representative of commercial interests, the Patriots would have to hold the British economy at risk to change the members' of Parliament behavior. This was accomplished not on land, but at sea.

Thus, this study offers a summary of land operations—a subject already described and analyzed in numerous books on the revolution—while delving into naval operations in greater detail. The reader should focus on the relationship between insurgent actions and strategy on land, and the complementary insurgent actions and strategy at sea. By holding Britain's commercial interests in the Caribbean, North America, and India at risk, and by demonstrating a threat of naval raids on the British Isles, the fledgling Continental Navy and American privateers—in league with the French—moved Parliament to consider whether principles of Parliamentary sovereignty was worth the loss of national wealth.

Abortive Invasion of Canada, 1775

The Patriots' first military campaign took aim at British Canada and sought to drive the British out of the province through the capture of Quebec City. Patriot leaders likewise hoped to entice French Canadians to join the cause, which might then provide diplomatic leverage over Paris as well. To effect this outcome, Brigadier General Richard Montgomery led an expedition north from Fort Ticonderoga, while Benedict Arnold led a parallel expedition from Cambridge, across the brutal wilderness, toward Quebec City (see Figure 6-3).

Case Studies in Insurgency and Revolutionary Warfare—The Patriot Insurgency

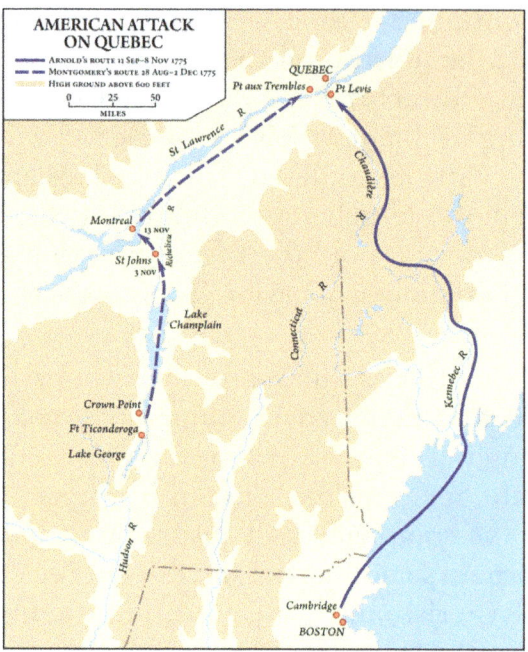

Figure 6-3: Patriot invasion of Canada, 1775.

Montgomery's troops besieged and captured Fort St. Johns in November, which led to British General Guy Carleton's decision to abandon Montreal and flee to Quebec City. Montgomery's capture of Montreal was a boon for the Patriot cause, but two problems emerged thereafter. First, the soldiers' enlistment contracts were set to expire at the end of 1775, and many of them refused to remain, preferring to return home rather than face prolonged hardship. Second, when Montgomery advanced with his depleted army for Quebec City, he left the incompetent General David Wooster in charge of Montreal. The political situation in and around Montreal was already challenging for the Patriots because the British Quebec Act had been favorably received by the French Catholic population. What could the Patriots offer in its place, other than idealistic words about liberty? Wooster's administration worsened the situation by bullying Loyalists and imprisoning elites who failed to come over to the cause. Likewise, the Patriots took supplies from the local farmers and paid with paper money instead of coin, which aggravated the French colonists.

Meanwhile, Benedict Arnold led just over a thousand troops north by sea from Cambridge, Massachusetts and up the Kennebec River. From there, the expedition struggled through swamps, forests, and bad weather to the Chaudierre River, which ran north to join the St. Lawrence River at Quebec City. By the time he reached the confluence, Arnold had only six hundred men left, and they were starving and without cannon.

Arnold's force joined with Montgomery, who had marched down the St. Lawrence from Montreal. On the last day of 1775, the combined force attacked Quebec City in a snowstorm and was bloodily repulsed. Montgomery was killed, Arnold was wounded, and Daniel Morgan—leader of an effective troop of riflemen—was captured. Arnold attempted to maintain a siege of the city, but he was eventually ordered back to Montreal. This failed expedition set the stage for Burgoyne's campaign in 1777.

The Campaigns in New England and the Middle States, 1776-1778

George Washington arrived at Boston as the new commander in chief in July 1775. He tried to instill discipline and impose some logical organization over the collection of volunteers and militiamen he had inherited. He confided in his letters to intimates that he was depressed and worried that the army would melt away from desertion, disease, and the expiration of enlistments at the end of the year. He argued repeatedly for an attack on Boston, but his officers advised against it. Finally, the stalemate was broken through the herculean efforts of Henry Knox, who led a party that hauled the artillery from Fort Ticonderoga across the Berkshire Mountains to Boston. In March 1776, Washington's troops seized Dorchester Heights in a surprise night move and trained their new artillery pieces on the city. This move made the British position untenable, and General William Howe withdrew to Halifax. The Continental Congress and Patriot networks throughout the colonies hailed the withdrawal as an impressive Patriot victory. However, Howe's army had been outmaneuvered, not defeated, and the pressing question was where it would land next.

In July, the British returned in massive force, landing at New York. The city was a major trade port, as well as a center of Loyalist support. Washington and the Congress concluded that their army was obligated to defend the colonies wherever the British attacked them. Their grasp of insurgent strategy had not yet matured, nor had their understanding of strategic feasibility. As a result, Washington attempted an ill-fated defense of the city despite the British naval strength, and at the Battle of Brooklyn in August, the Patriots were defeated and nearly destroyed. Washington pulled off a nearly miraculous retreat that saved his army but began a long, dispiriting campaign that ended with the demoralized Continental Army retreating across the Delaware River into Pennsylvania.

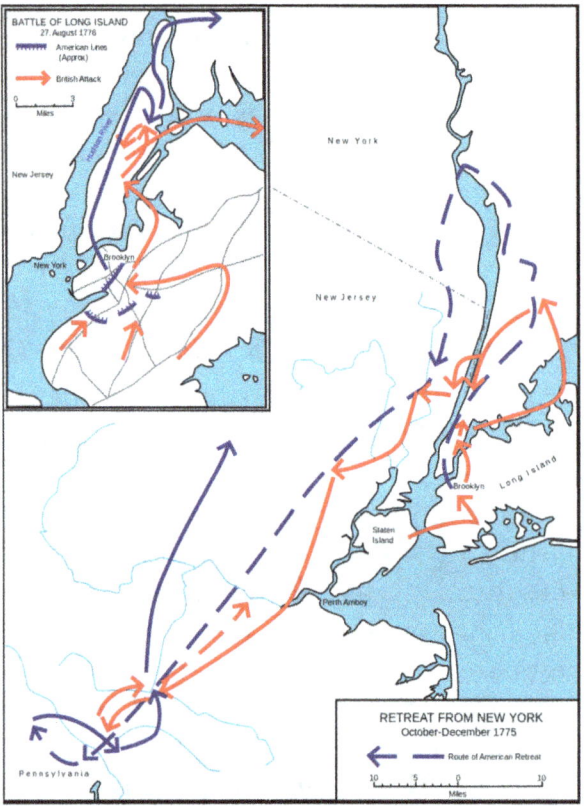

Figure 6-4: The Battle of Long Island, retreat, and counterattack, 1776.

With the onset of winter, the British forces, including Hessian mercenaries recruited in the German principalities, went into quarters to await the spring campaigning season. The Patriot cause was at its weakest state, and many wondered if the war might be lost. Washington chose this moment to strike in a brilliant winter campaign. In late December 1776, he led his army across the Delaware and surprised British forces first at Trenton and later at Princeton, inflicting two humiliating defeats on them. This surprising turn in fortunes heartened the Continental Congress and the Patriots throughout the country. This move demonstrated a growing sophistication in Washington's understanding of insurgent strategy. He correctly perceived the linkage among tactical victories, popular morale, political support, and gradual attrition of British will. After its surprising victories, the Continental Army went into winter quarters in Morristown, New Jersey.

The year 1777 was decisive. The British sent General John Burgoyne with an army from Quebec down the Richelieu River toward the Hudson River valley. Burgoyne's newly reinforced force, though boasting substantial numbers of men and cannon, were trained and equipped for European-style warfare. When faced instead with nearly impenetrable wilderness, they were unable to project combat power effectively down the Hudson Valley, and their progress south was slow and demoralizing. The campaign was poorly managed as well, and General Howe made the curious decision to depart New York City and seize Philadelphia, foregoing the opportunity to link up with Burgoyne to isolate New England. Instead, Burgoyne's slow-moving army fought a pair of engagements at Bemis Heights and Freeman's Farm in New York and eventually surrendered at Saratoga. This stunning defeat shocked the British government and impressed the French with the Patriots' surprising resilience. It led the following year to a treaty of alliance between the United States and France. As the war against Britain widened, Spain and the Netherlands also allied with France.

Meanwhile, General Howe's army departed New York City and landed in Maryland, intent on taking the Patriot capital, Philadelphia. As Howe marched north, Washington formed his army along Brandywine Creek to stop him. Instead, on September 11, 1777, Howe's redcoats marched around the Patriot right flank, crossed the creek, and defeated Washington's forces. The Contental Army escaped destruction, but Howe captured Philadelphia two weeks later, forcing the Congress to flee to Lancaster, Pennsylvania, and later to York.

Following his seizure of the capital, General Howe moved west toward the Patriot base at Reading. Washington decided to intercept him at Germantown, but his complex plan to attack with four separate columns faltered in the foggy weather and confusion of battle. The British thus inflicted yet another defeat on the Continental Army, but the French were nevertheless impressed by Washington's audacity and the persistence of the Patriot troops. Despite a string of tactical defeats, the Patriot soldiers and leaders demonstrated pluck and cohesion. This observation contributed to the French decision to join the war.[29]

During the harsh winter of 1777–1778, the Continental Army camped at Valley Forge, Pennsylvania. Despite lack of food, shoes, shelter, and ammunition, the Patriot soldiers endured under the leadership of Washington and his generals. Washington employed the Prussian émigré officer Friedrich von Steuben to drill his soldiers, increasing their discipline and battlefield resilience. Their growing toughness and competence was demonstrated at the Battle of Monmouth in June 1778, when the Continental Army attempted to attack the rear of the British army that was withdrawing from Philadelphia toward New York. British General Cornwallis struck at the disorganized Patriot forces under General Charles Lee, seizing a key hedgerow. Washington arrived on the battlefield and organized a counterattack that drove Cornwallis back. By the end of the engagement, the Patriot forces fought their British counterparts to a draw—the first time Patriot regulars fought on equal footing with the redcoats. The Battle of Monmouth was the last major battle in the northern colonies.

This initial phase of the Revolutionary War tested the political and military strength of the Patriot insurgency. Its political network functioned reasonably well, keeping everyone informed of military developments and generating moderate support for the Continental Army. The army's logistical foundation, however, remained poor. Part of the problem was the Congress' lack of power to tax the colonies to generate revenue for purchasing uniforms, ammunition, arms, food, and transportation. Washington continued to complain about the lack of support, but he and his officers did remarkably well with the small support they received. Washington and his better officers—the Marquis de Lafayette, Nathaniel Greene, Anthony Wayne, "Light Horse" Harry Lee, and arguably William Irving and John Sullivan—demonstrated a growing capacity to use the terrain to their advantage, push their troops to heroic exertion, and maneuver competently against professional British forces.

The northern campaigns also saw the Patriots striking back at the Indian tribes cooperating with the British. Under Mohawk leader Joseph Brant, tribes of the Iroquoi Confederation conducted devastating raids against the western portions of Pennsylvania and New York. In 1779, Washington directed John Sullivan to lead an extended expedition against them. The Patriot forces destroyed native villages.

After Monmouth and the entry of France into the war, the British strategy in the north switched to limited raiding in an attempt to draw

Washington and the Continental Army out into terrain where they could be defeated. Washington refused to comply, keeping the Army out of reach in fortified positions around New York. In July 1779, British Major General William Tryon led an extended raid along the coast of Connecticut, destroying military and public buildings and stores. The raiders also destroyed private homes and churches in an effort to goad Washington to respond. Instead, local militias turned out, but they were unable to defend against Tryon's professional soldiers. The Patriots absorbed the British raids without resorting to the major battle that the British wanted, knowing that time was on their side.

As British major operations ceased in the north, the Continental Army continued to harass and ambush the British—most notably by attacking isolated outposts at Stony Point (July 1779) and Paulus Hook (August 1779). Small, well-trained, and well-led Patriot columns surprised and captured British detachments. These victories helped Washington to bleed the British army as well as maintain Patriot influence in New Jersey and New York. With the failure of Britain's northern campaigns, the war was about to shift south.

The Southern Campaigns, 1778-1783

The British decided to switch their focus toward the southern states because they anticipated stronger Loyalist support there, and they hoped to entice black slaves to revolt against their masters, thus disabling the southern economy. After recruiting strong Loyalist forces, the British would lead them north through the Carolinas and Virginia to secure the areas the British Regulars and Provincials liberated. In late 1778, they seized Savannah, and in May 1780, they also captured Charles Town, South Carolina, along with its large garrison of Continental soldiers. The character of the subsequent southern campaigns differed from that of the earlier northern phases because of the rough terrain of the hinterland, where most of the mobile actions played out, and the prominence of Patriot and Loyalist militias. At the abortive Battle of Waxhaws, a detachment from the Continental Army, reinforced with local militiamen, was massacred by British and Loyalists commanded by Banastre Tarleton after attempting to surrender. Patriot propagandists used this incident after the fact to vilify the British and encourage recruitment, but it underscored the hatred between Patriots and Loyalists.

Under Sir Henry Clinton, the British followed up the seizure of Charles Town with campaigns in the back country of South Carolina. Clinton then departed for New York, leaving General Charles Cornwallis to complete the subjugation of the southern colonies. His campaign scored a major victory at Camden, South Carolina in August 1780. General Horatio Gates—the hero of Saratoga—took command of a mix of Continental Army regulars and untried Virginia and North Carolina militias. He intended to push the British out of Camden, but in the ensuing battle, the outnumbered but better trained British regulars smashed the Patriot force, inflicting nearly two thousand casualties with little cost to themselves.

Following the capitulation of Charles Town, there were dozens of skirmishes between Loyalist and Patriot forces. Most of the clashes resulted in Patriot victories, and they disrupted British plans long enough to allow Nathaneal Greene to rebuild his forces.

At King's Mountain, North Carolina, Patriot militias surrounded and defeated Loyalist militia in October 1780. Anger and a desire for revenge of the Waxhaw massacre led the Patriots to kill and wound Loyalists trying to surrender. The victory boosted the flagging Patriot morale, but subsequent abuse and execution of prisoners revealed the increasingly brutal conflict between American Patriots and American Loyalists. The animosity between the two sides would make postwar political settlement difficult. However, with the threat of Patriot militias in the mountains of North Carolina, General Cornwallis abandoned his plan to invade the colony and instead returned to South Carolina. In that same month, Washington assigned Nathaniel Greene to command in the south.

Greene decided not to seek a conventional battle but instead split his forces in North Carolina to raid, gather supplies, and boost the morale of Patriot supporters. He sent General Daniel Morgan west with a detachment of regulars and ordered him to join with Patriot militia west of the Catawba River. Alerted to the presence of Morgan's army, Cornwallis sent Banastre Tarleton to intercept and destroy it. The British detachment first hastened to the key fort of Ninety-Six, but upon learning that Morgan was not there, Tarleton, reinforced with more British regulars, pushed north toward the Broad River. Morgan retreated toward the river, hoping to cross it before Tarleton could close, but when he judged that he did not have time to cross, Morgan instead directed his troops to form for battle at Cowpens. On January 17, 1781,

Tarleton's forces arrived exhausted, malnourished, and overconfident. Morgan carefully arranged his forces into three lines: sharpshooters, militia, and regulars. His plan was to use the first two lines to fire at the advancing British regulars and then withdraw, so as to prevent them fleeing and to encourage the British attackers to continue. The third and final line of regulars, entrenched on a rise, would then hold and complete the defeat of the British. His plan worked perfectly, and the Patriots inflicted a decisive defeat on Tarleton's forces, nearly wiping them out completely.

Cornwallis decided he must destroy Nathaniel Greene's army and commenced a campaign northward, chasing Greene all the way to Virginia. On March 14, 1781, Greene's army was in position at Guilford Courthouse near present-day Greensboro, North Carolina. Though outnumbered, Cornwallis decided to attack. Greene's forces mauled the attacking redcoats, killing and wounding about a quarter of Cornwallis' force. The Patriot army eventually broke off the engagement and retreated, leaving the battlefield in British hands. Cornwallis thereupon called the battle a British victory, but political opponents back in the British Parliament saw it as a Pyrrhic victory. Charles James Fox famously lamented, "Another such victory would ruin the British Army!" After the battle, Cornwallis withdrew to the Carolina coast to refit his army with supplies brought by the Royal Navy. When Cornwallis decided to move north in an attempt to pacify Virginia, Greene marched south to liberate South Carolina and Georgia .

The British army under Cornwallis spent the summer raiding deer into Virginia before he reached Yorktown, Virginia in September. Pursuant to the confusing orders he received from General Clinton, he dug in and began to fortify the deep-water port. Meanwhile, the French and the Patriots conceived a plan for a joint campaign in which French Admiral de Grasse's fleet, newly arrived from the Caribbean, would blockade Cornwallis while the combined French-American army under Washington and French General Rochambeau closed on Yorktown. The allied army departed from the vicinity of New York City in the summer of 1781 and marched southward toward Cornwallis' position. At the naval battle of Chesapeake Bay on September 5, 1781, Admiral de Grasse defeated British Admiral Thomas Graves, preventing the British from evacuating or reinforcing Cornwallis. Washington and Rochambeau arrived with their allied armies and besieged the British. The resulting siege lasted from September through October 19th, when the

British finally surrendered. This was the last major land battle of the war, and it led to the British government opening negotiations to end the war. The 1783 Treaty of Paris concluded the conflict, and American independence was intact.

Analysis of Land Operations

The Patriot insurgency started with a collection of ill-equipped, poorly trained local militias in 1775 and eventually built the Continental Army that could fight with British regulars. Washington and his lieutenants fought and learned as the campaign in the northern colonies progressed, punctuated with many humiliating defeats and a few morale-boosting victories. Washington proved himself an adaptive strategist as he kept his forces from being decisively defeated while prosecuting a series of raids, feints, and occasional battles. He was successful in convincing the British commanders that they could not hope for a military victory in the north. When they decided instead to pursue a political and military victory in the southern colonies, Washington's subordinates—principally Greene and Morgan—led an irregular warfare campaign, occasionally accepting conventional battle to bleed the British forces. Throughout the campaigns, the Patriot leaders sought to maintain political support for the revolution while coercing Loyalists, often with brutal tactics. In the end, Washington, with decisive aid from the French army and navy, trapped and defeated the major British force, leading to the British government's decision to abandon its aims in North America.

Naval Operations

The Patriot Insurgency at Sea

Insurgents are effective when they recognize their strength and bring it to bear against the strong power's weakness. The Patriot insurgency at sea capitalized on a strong maritime tradition and deployed ships and men who made their living at sea, against their British counterparts. It was an essential element to the United States garnering independence and recognition from the United Kingdom. More men served at sea than in the land campaigns, more men were taken prisoner, and more damage done to the British economy.[30] The war's naval component had profound impact at the tactical, operational, and strategic levels of warfare. Waterborne craft were a tactical imperative in a nation that relied on rivers, lakes, and littoral ocean environs more than roads for transportation. Ships provided supporting fires, protected harbors, and moved men and materiel. The Patriot insurgency at sea served as an operational-level enabler for colonial victory ashore.[31] The maritime campaign had strategic impact because it directly attacked British commerce, damaged the British economy, and hurt the British population, the enemy's center of gravity. That the fight was in their territorial waters and in a few occasions landed on their shores sowed fear and affected their collective psyche. Together these drained the population's willingness to continue the fight. Additionally, while victory at Saratoga provided the confidence needed for France and Spain to join the conflict, they identified with the Patriot insurgency first because American sailors brought ships, supplies, and sailors from their longstanding nemesis into their ports as prizes and prisoners. When the alliance formed, the conflict became a world war between strong powers. The fight then expanded primarily at sea in the West Indies, the English Channel, the northern Atlantic, and even the Mediterranean.

During the age of sail, there were two forms of naval warfare. If a nation state possessed a robust Navy, many ships that were fast and well armed, they would often engage in *guerre de escadre,* or "war of the squadron." This described force-on-force battles wherein ships would meet, maneuver to advantage, and fight with canon and muskets at range then often close for boarding and hand-to-hand combat with a mix of small arms, cutlasses, and knives. The Royal Navy was capable of this form of combat, as were France and Spain. These countries had large men of war, officers, and crew that spent careers training or conducting war at sea, could form fleets and flotillas, and maneuver them together to provide supporting fires.

The colonies could not even muster a frigate at the beginning of the war, and they never outfitted enough ships to form more than a squadron of five or six ships to sail in unison. For a lesser naval power with a small and inexperienced fleet *guerre de course* or "war of the chase" was a more attractive strategy. In this approach, a vessel, often sailing independently, would search for an enemy's merchant shipping. Then, they would pursue and fire on the enemy ship until the merchantman "struck,"—in other words, lowered their national flag in a sign of surrender. Its crew would become prisoners, and the vessel taken over by a "prize crew," a subset of the victorious ship who would sail it to the closest friendly port and sell the ship and its goods. Attacking logistics and raiding commerce hampered a state's ability to sustain an expeditionary force and interrupted its economy.

Navies are costly to build and maintain. As a result, governments enlisted the services of commercial interests willing to conduct *guerre de course* on their behalf. The governments would issue letters of marque and reprisal, legal documents that provided legitimacy, preventing the bearers from being tried under international law as pirates. During the Seven Years' War, there were at least three hundred privateers sailing with the authority of the Massachusetts colonial government.[32] The Continental Congress issued 1,697 letters of marque, and the colonies authorized even more.[33] Pre-printed letters of marque and reprisal that needed only the name of the captain and his command to be added sped the legal process of joining the cause.[34] Benjamin Franklin and Silas Deane, congressional liaisons in France, bought ships, issued commissions to commanding officers, and supported privateering in a form that sometimes made it unclear if the ship were of the Continental Navy or a privateer. Even with clear legal documentation, privateers were not recognized by the British crown. Parliament passed a piracy act that stipulated Americans captured on the high seas could be tried for either piracy, treason, or both. Placed in these categories, they could not receive the privileges granted to prisoners of war. For example, they could not be released in a prisoner exchange nor given parole.[35] Many American sailors chose to sail as privateers despite the danger of being tried as pirates. Throughout the war, the Continental Navy had a maximum of sixty-four ships, while the number of licensed privateers rose to 1,697.[36] Those who chose to sail on ships of the Continental Navy obtained a lesser share of any captured vessel; Congress appropriated a share of the prize money to support the war.

The American practice of *guerre de course* proved successful for a time. The British lost 2,980 merchants, compared to the American loss of 1,351 with a cost delta of $50 million.[37] Insurance rates for British shipping rose to 23 percent of the cargo's value, and during the war, merchants experienced a loss of £66 for every £100 when compared to figures before the war.[38] Many British businessmen appealed to Parliament to end the war due to their losses, and the impact was felt by all subjects of the crown.

Edgar Maclay describes this impact:

> A careful review of British newspapers, periodicals, speeches in Parliament, and public addresses for the periods covered by these two wars (American War of Independence and War of 1812) will show that our land forces, in the estimation of the British, played a very insignificant part, while our sea forces were constantly in their minds when "the American war" was under discussion. When England determined to coerce the refractory Americans, she little thought that she was inviting danger to her own doors....But had they anticipated that American cruisers and privateers would cross the Atlantic and throw their coasts into continual alarm; that their shipping, even in their own harbors, would be in danger; that it would be unsafe for peers of the realm to remain at their country seats; that British commerce would almost be annihilated; that sixteen thousand seamen and eight hundred vessels would be taken from them – they would have entered upon a coercive policy with far greater hesitancy.[39]

Striking at British coffers, *guerre de course* thus eroded British will. American maritime forces were adept at this approach due to a strong maritime tradition that included commerce and smuggling. Many seafarers had been running from the Royal Navy for years.

When France, Spain, and other European powers joined the rebelling Americans, the Royal Navy found itself engaged in *guerre de escadre*. As a result, the conflict at sea felt like a battle against insurgents while trying to protect merchant and logistic shipping, but then grew into a global war between nation states. The conflict at sea began as an insurrection of smugglers but ended as a fight between strong powers.

Maritime combatants joined the Patriot insurgency under a variety of diverse banners. Just as colonies were protected by militias ashore, most formed navies to protect their interests at sea. Of the thirteen colonies that voted for independence, only New Jersey and Delaware did not form a navy. Similarly, just as the Continental Congress raised an army, it established a navy. Additionally, using his authority as commander in chief of the Continental Army, George Washington hired several ships to attack the British and provide his forces with logistical support. This element is commonly referred to as "Washington's Navy." Finally, the bulk of the maritime war, and the largest part of the Patriot insurgency, were conducted by privateers granted legal authority through the letters of marque and reprisal issued by both the individual colonies and the Continental Congress. Each of these approaches to the Patriot insurgency at sea was relatively successful in achieving their aims because the colonies had a large population of sailors and officers who were tired of British oppression at sea.

The Gaspee Affair

To administer the Navigation Acts that directed the colonies to trade only with England, the British government purchased additional vessels to serve as custom enforcement vessels. Officers in the Royal Navy held two commissions, one in the Royal Navy and one from the British Treasury.[40] As a result, these officers and their crews had the authority to hail, stop, and search American merchants to ensure they were complying with the law. The Royal Navy had an extra incentive to be aggressive because, in a practice similar to privateering, they were authorized to keep a percentage of seized goods.[41] Rhode Island was a haven for smugglers, so Lieutenant William Dudingston of HMS *Gaspee* patrolled Narragansett Bay, putting a damper on illicit trade. Dudingston was particularly aggressive and reportedly bullied colonials.[42] In March 1772, Rhode Island merchant John Brown generated a petition and presented it to Chief Justice Stephen Hopkins who provided a legal opinion that stated all Royal Navy officers must present themselves to the local governor before conducting any business in the colony's waters.[43] Dudingston refused and indeed threatened to hang any man who tried to oppose the *Gaspee*.[44]

On June 9, 1772, while chasing an American merchant named *Hannah*, Dudingston fell victim to the fact that he was sailing in unfamiliar waters. *Hannah* raced purposefully into shallow waters; *Gaspee* followed and ran aground. Now the ship had to wait for the tide to come in before she could sail off the shoal. Fed up with Dudingston's harassment, during the night, a group of Rhode Islanders led by Abraham Whipple rowed out to her under the cover of darkness and attacked her. During the battle, she was set on fire, which likely spread to her magazine, and the ship exploded.[45] King George offered a reward for the capture of these Rhode Islanders, but no one was ever turned in.

Chapter 6. Patriot Insurgency

The First Lake Champlain Maritime Campaign, 1775

Colonial America relied on its waterways as a primary transportation pathway; it was easier for commerce to move by boat than by cart or carriage. This meant, just as it is true today, most goods spent some time afloat whether in the open ocean or upon navigable rivers and lakes. These routes were also the most efficient means to move troops and controlling the region demanded protecting key points with a fortress and garrisoned troops. One important inland corridor connected Canada to New York. From Quebec, one could travel on the St. Lawrence River, to the Richelieu and Lake Champlain followed by a short overland jaunt (later connected via canal) to the Hudson River, which flows to New York Harbor and the Atlantic Ocean. If the British controlled this avenue, they would cut the colonies in two, dividing New England from the mid-Atlantic and southern colonies. Similarly, if the Patriot insurgency controlled this corridor, they would remain geographically united and could move freely from the Atlantic to Canada.

After Lexington and Concord, British forces held Boston but were under a colonial siege. They were able to resupply from the sea. Several military-minded colonials saw the opportunity to attack British interests in Canada. A businessman and merchant sailor named Benedict Arnold who served in the Connecticut militia proposed to the Massachusetts Committee of Safety to take Fort Ticonderoga. They concurred with his plan and provided him with a commission. En route, he met Ethan Allen and the Green Mountain Boys, a Vermont militia that also decided to attack the British position at Fort Ticonderoga. It was captured easily on May 10, 1775. Though Arnold did not command any troops himself, his experience as a sailor proved invaluable. He realized that to control Lake Champlain, one had to seize the forts that protected the river but also form a brown-water navy to control its waters.[46] Arnold took command of a vessel owned by a local landowner, naming it *Liberty*. On May 14, Arnold sailed north with a force aboard *Liberty* and two bateaux. Continuing up the Richelieu River, on May 16 and 17, 1775, they attacked St. John's, capturing HMS *George*.[47] Arnold dubbed her "*Enterprise*."[48] From the British perspective, these were remote outposts and as such did not require robust garrisons. Nevertheless, this American advance meant Lake Champlain was now devoid of British strongholds and vessels. For the British to control this region, they now had to win it back, and to retake the lake would mean they now also had to reconstitute a navy. Conversely, Arnold's actions meant the colonials were poised to invade Canada. Thus, this would not be Arnold's last campaign on the lake; his skill as a seaman would be important to colonial victory.

Washington's Navy, State Privateers

While Benedict Arnold pushed north up Lake Champlain, George Washington focused on the Boston siege. Knowing the British logistic train would serve as a weak point and prizes could support his own force, Washington formed a fleet to conduct *guerre de course*. On September 2, 1775, he commissioned *Hannah* (not the same Hannah the *Gaspee* pursued in 1772) under the command of Nicholson Broughton. Her first prize was *Unity*, a vessel carrying naval stores. Eventually, Washington's Navy included six schooners and a brigantine. Their greatest prize was *Nancy*, a ship carrying two thousand muskets, thirty-one tons of musket shot, three thousand round shot, several barrels of powder, and a thirteen-inch brass mortar.[49] In total, Washington's Navy captured thirty-eight prizes, totaling $600,000. This was far more valuable as war materiel transferred from the British to the colonials, the moral victory it provided to the revolutionaries, and the fact that it significantly encroached upon the British ability to sustain their forces in Boston.[50]

An interesting mix of state-funded navies and privateers joined Washington's Navy in engaging the British. All of them sought a combination of self-governance and profit, though each placed emphasis on freedom versus finances to varying degrees. While at sea, however, their tactical goals were the same. As a result, state navies and privateers were able to combine forces, albeit loosely, against the Royal Navy. For example, on June 17, 1776, Connecticut State cruiser *Defense* heard cannon fire. Upon investigating the source, *Defense* met four schooners attacking two British transports. One was the Massachusetts State cruiser *Lee* and three privateers. Together they pursued the Royal Naval vessels into Narragansett Bay and engaged the Brits who eventually struck. Three hundred British soldiers were captured.[51] Agents in ports throughout the Americas and Europe conducted the purchase, resale, and payment to the victorious captain and crew. Like the privateers, these capitalists made a fortune and with much less personal risk. For example, William Bingham acted as an ambassador for the Continental Congress and a prize agent at the Caribbean Island of Martinique.

With the fortune he amassed while serving in this role, Bingham established the first bank in America.[52]

All of these efforts were promising. Still, if the colonials were to defeat the British the Continental Congress, they needed a major naval force. Doing so would provide a modicum of legitimacy to the effort both in North America and abroad. The Continentals needed a navy.

Formation of a Naval Committee/Marine Committee

During the Continental Congress session in the fall of 1775, John Adams proposed forming a Navy. Most representatives initially thought this ill advised. Establishing a navy would be seen as more belligerent an act than forming a militia, one that blatantly crossed the threshold of rebellion at a time when there was still hope that American grievances may be addressed through political negotiation. Additionally, the Royal Navy seemed an undefeatable foe.[53] Unable to garner support, the matter rested until Captain John Barry returned from England aboard *The Black Prince*. He brought with him British newspapers reporting the Royal Navy ordered eight men of war to set sail for North America.[54] On October 13, 1775, Congress authorized a single ship. Today, this event is marked as the official birth of the American Navy. Like many other merchant captains, John Barry wanted to return to sea in support of the cause. However, he was initially assigned to outfit ships for war.[55] He set about converting *Black Prince*, which became the Continental Navy's ship *Alfred*. Barry directed work on *Montgomery* and was then asked to oversee the action in four Philadelphia slipways where timber, line, and canvas became four of the Continental Navy's frigates.[56] Finally, he was given command of *Lexington*.

During the conflict, management of the Navy took many forms. The original Naval Committee grew from seven men to a Marine Committee of thirteen in December 1775. In December 1779, a Board of Admiralty replaced this with two congressmen and three appointed citizens. Finally, in 1781, initially Alexander McDougall and later Robert Morris became the sole administer (not unlike a Secretary of the Navy) with subordinate boards in Boston and Philadelphia to manage tasks from operations to distribution of prizes.[57] The original naval committee comprising Silas Deane of Connecticut, John Langdon of New Hampshire, and Christopher Gadsden of South Carolina were soon joined by John Adams of Massachusetts, Richard Henry Lee of Virginia, and Stephen Hopkins and Joseph Hewes of North Carolina. They directed the navy's administrative and operational aspects, delineating its uniform, regulations, and how captured vessels would be parceled into prizes for the victorious crew.

Authorization of a single ship soon turned into a squadron of converted merchants. These included *Alfred*-24 and *Columbus*-20, outfitted as frigates, *Andrea Doria*-14 and *Cabot*-14 as brigs with six pounders, *Providence*-12, *Hornet*-10, and schooners *Wasp*-8 and *Fly*-8.[58] Then in December 1775, Congress authorized construction of thirteen frigates to be built in shipyards spread through seven colonies.[59] A frigate was a purpose built war ship with three square-rigged sails but with only one gun deck compared to the larger, slower men of war that would include two more. Five of the frigates would be of thirty-two guns, *Hancock, Raleigh, Randolph, Warren,* and *Washington;* five of twenty-eight, *Effingham, Montgomery, Providence, Trumble,* and *Virginia;* and three of twenty-four, *Boston, Congress,* and *Delaware*.[60] Some of these enjoyed successful careers; some were lost before they put to sea. Most salient, however, was that even before the formal declaration of independence, this act of building a navy suggested that Congress represented legitimacy, or at least an attempt to assert sovereignty.

Esek Hopkins was appointed as the fleets' first commodore. His first order was to attack the British fleet in the Chesapeake Bay. A privateer from the Seven Years' War, Hopkins was not confident that his command was prepared to face the Royal Navy. Exploiting a clause in his orders that gave him authority to exercise judgment, he sailed instead to Nassau in the Bahamas. Shortly after getting underway, *Hornet* and *Fly* became separated from the rest of the squadron and were not seen for the rest of the voyage.[61] Arriving off Nassau on March 3, 1776, Hopkins' flotilla captured a British garrison and seized over one hundred cannon and other war materiel, which would prove valuable to the cause.[62] Governor Brown was taken prisoner and later exchanged for Lord Stirling, a general in Washington's army.[63] Taking high-ranking officials and commanding officers as hostages to be paroled or exchanged was a common practice of the day. On April 4, 1776, the squadron captured *Hawke*-6, and the next day, *Bolton*-12.

Hopkins wisely avoided confrontation with a superior force, attacked a British garrison capturing war materiel that would benefit George Washington's campaign in New York and holding British officials as hostage, took two ships that could either be sold or employed by the Continental Navy, and returned to homeport without significant losses. For a first deployment, one cannot imagine a better outcome. His approach was not unlike Benedict Arnold's campaign into Canada. Though leading a force representing an insurgency at sea, he chose targets with a high chance of success. Despite all of this, Hopkins was censured for not following orders, his enemies claiming that he did not "annoy the enemy's ships upon the coasts of the Southern States."[64]

External Support

The Patriot insurgency needed gunpowder and other war material to fight at land and at sea. Anticipating rebellion, the British secured and protected military supplies. Colonists were able to seize about eighty thousand pounds of powder.[65] Then, the colonial insurgency at sea was able to capture guns, ammunition, and powder, but more was needed to sustain the effort. External support was developed through European powers who were rivals of the British and sympathetic to the American cause. Smugglers would bring the arms and equipment from France, Spain, the Netherlands, and Russia. Britain handled this challenge coyly, carefully. Upsetting these rivals might draw them into the war on the side of the American insurgents. On December 22, 1775, Britain passed the Capture Act. This formally closed American waters to trade and announced the Royal Navy would conduct visit, board, and search operations on ships within their territorial waters to ensure illicit cargo were not headed to colonies.[66] The insurgency at sea sailed around the Capture Act legally by obtaining French papers or even French officers for ships carrying war materiel to the United States. The British, recognizing this ruse but, remaining consistent with an approach that was rightly concerned about France formally entering the war on the side of the colonials, decided they would not impede French ships or vessels.[67] Because their efforts to block these logistic lines of operation were ineffective, fully 90 percent of the American powder used in the first two years of the war arrived by sea.[68]

The Second Lake Champlain Campaign, 1776

Benedict Arnold, Ethan Allen, and the Green Mountain Boys defeated the British garrisons along Lake Champlain and the Richelieu River. Then, British forces in Canada redeployed to reinforce Boston. Benedict Arnold and others lobbied for an invasion of Canada. Arnold received a commission as a colonel in the Continental Army and was assigned troops to march north through what is now the state of Maine.

General Philip Schuyler would sail north on Lake Champlain, then march northwest, and the two would meet in Quebec with the intention of taking it, driving the British out of Canada, and perhaps drawing the Canadians into the war against Britain. General Richard Montgomery relieved Schuyler when he fell ill. After an arduous trip by Arnold and his force, the two elements did rendezvous and attack Quebec on New Year's Eve 1775. Montgomery was killed; Arnold was wounded.[69]

This arduous campaign did not achieve its intended strategic goals, but it did have a positive strategic effect of setting the stage for a robust defense against the British in 1776. The maritime operations of this first campaign focused mostly on transporting land forces and supporting them in both maneuver and attack on inland waters. However, the naval war on inland waters was just beginning, and it would remain an important element throughout the conflict.

The British developed a strategy for 1776 that would divide the colonies in two. One element would capture New York City, maintaining a presence there to ensure continued use of its valuable harbor for resupply from the sea. From this anchor point, they would advance north on the Hudson River. A second element would form in Canada and invade the colonies by heading south on the Richelieu River and Lake Champlain. When the two elements met, a line would form separating the New England from the rest of the colonies.[70]

Both the British and Continental forces built boats and small ships on Lake Champlain in anticipation of another fighting season. The British under Sir Guy Carleton still planned to maneuver south, using the Richelieu River, Lake Champlain, and the Hudson River to as aquatic highways to connect with General Howe in New York. The American forces in the region, now under the command of Benedict Arnold, were directed by General Gates to fight a defensive action against any invasion. Arnold had a flotilla that consisted of a sloop, three schooners, eight gondolas, and four galleys totaling ninety-four guns and seven hundred men.[71] Similar to the Continental Army throughout the war, Arnold's squadron, while diminutive, was a "fleet in being." It served as a known adversary that simply by its presence forced the British to pause and develop an effective means to defeat it.[72]

To meet this threat, the British built a fleet at their position on the northern side of Lake Champlain. Their squadron was similar in size, twenty vessels, armed with a similar number of cannon, but of higher weight. As a result, they could lob heavier, more destructive rounds at the Americans.[73] By October 1776, their force on Lake Champlain included "a ship-rigged sloop, two schooners, a gondola, an enormous radeau, 20 gunboats, 24 unarmed longboats, 24 armed longboats, and 450 bateaux to ferry troops."[74] Constructed on the lake, the sloop, HMS *Inflexible*-18, weighed 180 tons and alone could destroy the American flotilla.[75]

The Battle of Valcour Island

These two elements would eventually clash on October 11, 1776 in what is recognized as the first battle in US Navy history, the battle of Valcour Island. Valcour Island lies on the southern portion of the lake, close to the western shore. Benedict Arnold moored his ships in the channel between the two in a crescent moon shape, with its apex to the north and all of the vessels pointing to the south. The British force sailed past the island, thereby finding them exposed with the Continental Navy astern of them. A battle ensued that found the American forces garnering some success through surprise and superior positioning to begin the battle. HMS *Carleton* (presumably named for the governor) was put out of commission. Arnold directed much of the gunfire from his command ship, *Congress*, literally aiming the canon himself. Native Americans supported Arnold by providing harassing fire from shore. When the Brits maneuvered out of range, Arnold grabbed the opportunity to retreat with his forces. As before, he suffered a tactical defeat while obtaining yet again a strategic gain. The British were unable to use Lake Champlain as an invasion route for a second year.[76]

Sustaining Foreign Assistance and Privateering

The Continental Congress knew that *guerre de course* was the right strategy to employ against the British. It was also well known that chance of outlasting the British would be far greater with the support of its French and Spanish adversaries. Therefore, Congress sent emissaries to France. Through Silas Deane, and later Benjamin Franklin, the colonies obtained funds, munitions, and even ships. Franklin in particular was very popular in the French court. He arrived aboard *Reprisal*-18, commanded by Lambert Wickes in the fall of 1776.[77] They even captured prizes en route and took them into Nantes. The British protested, but Wickes continued. He captured five more prizes in the Bay of Biscay in the beginning of 1777 and then led a squadron of three ships through the English Channel and the Irish Sea, capturing another eighteen ships.[78] With these actions, the Patriot insurgency at-sea strategy began to work. British ships now would only sail in convoy, and the insurance rates began to rise.

Another captain, Gustavous Conygham commanded an American merchant vessel sailing for Amsterdam to obtain munitions. His ship was captured by the Royal Navy, but he and his crew were able to take it back before arriving in England. When he arrived at his destination, British officials prevented the Americans from trading. Conygham sold the ship and began to look for work.[79] He was recommended to Benjamin Franklin who provided him with a commission and a ship. "The appointment proved problematic. Though it placed Conygham in government service, his ship and crew were 50 percent financed, due to the mission's chronic lack of funds, with private money."[80] Conyngham was a Continental officer, commanding a privateer. He took command of *Surprise* and began taking prizes.

On May 3, 1777, his second day underway on *Surprise*, Conyngham and the small crew took *Prince of Orange*, a British royal packet sailing between the Netherlands and England that he suspected had British gold onboard. When he brought the prize into Dunkirk, *Prince of Orange* was released, *Surprise* was confiscated, and Conyngham was thrown in jail, all to appease the British.[81] The French were still pretending to be neutral. Conyngham's voyage was a personal failure, but a strategic success. The insurance rates for commerce in the English Channel increased by 10 percent at the threat of privateers were inflated.[82]

The American delegation in France played politics, condemning Conyngham as misguided while obtaining and outfitting a ship in anticipation of his release; his new vessel was dubbed *Revenge*, and like *Surprise*, it was privately financed.[83] With *Revenge*, Conyngham captured twenty prizes in fourteen months. Some he ransomed, demanding payment for release while holding the commanding officer hostage until remitted. Others were taken into Spanish ports to be sold off.[84] Later the Spanish were angered when he took a Spanish ship ferrying British cargo. He ultimately decided to leave European waters—the politics were too troublesome—and continue privateering in the Americas.

In February 1779, Conyngham had to respond to a lawsuit filed by some of his prize crew in Philadelphia. They accused him of concealing funds earned from prizes. The fact that *Revenge* was privately owned emerged as an important issue. Conyngham claimed he did not know the vessel was a privateer, rather than a Continental naval vessel, and he never followed orders of a private owner.[85] Congress settled the matter by dropping the matter against Conyngham, but selling *Revenge*. The ship ended in private hands and then leased to Pennsylvania as a privateer and under Conyngham's command again.[86] Unfortunately, he was captured by the Royal Navy and found himself in jail again, first in New York and then in Mill Prison in London. He dug his way out in November 1779, escaping to Amsterdam.[87] Sailing home, he was recaptured in 1780 and placed back in Mill Prison from where he escaped yet again. Conyngham continued to fight for his prize money in the American court system while declaring his status as an officer of the Continental Navy. The commission Benjamin Franklin provided him was scoffed as "temporary." He never garnered the recognition or the restitution he deserved.

Father of the American Navy

Naval historians debate which American Revolutionary War figure should receive recognition as "Father of the American Navy." Some suggest that John Barry's contributions, starting with his role in directing the conversion of many merchant ships—literally *building* the American Navy—followed by several successful cruises at sea earns him this reputation. Others lament that the most successful commanders of the war, like Lambert Wickes, perished during the war and thus were not as easy to recognize. John Paul Jones eventually garnered this title due to his success at sea, a willingness to self-promote, that President Theodore Roosevelt's sought an iconic hero to bolster the US Navy, and that Jones' alcohol-preserved remains were discovered in France. An elaborate sarcophagus under the Naval Academy Chapel became his final resting place where midshipmen and young naval officers could perform a nautical pilgrimage. Generations that visited under the serious gaze of an ever-present honor guard would quip, "Jones was so ornery that even in death, it takes a church to keep him down, a bath of alcohol to keep him happy, and two Marines to keep him quiet."

Born in Scotland in 1747, John Paul Jones decided at a young age to go to sea to seek his fortune. He became a merchant sea captain but encountered some legal difficulty when it was suspected that one of his crewmembers died as a result of harsh discipline.[88] The details are lost to history and conjecture, but Jones decided to change his name, adopting "Jones" as his surname, and depart commerce in the West Indies for America. He arrived as revolution was fomenting and volunteered to join the US Navy as one of its first officers. His first assignment was as first lieutenant aboard Commodore Hopkins flagship *Alfred*-24.[89] Therefore, he was aboard for its maiden voyage to the Bahamas. When the commanding officer of *Providence*-12 was court martialed, Jones took command on May 10, 1776.[90] He cruised up and down the east coast, ferrying troops between Rhode Island and New York. When directed to provide escort to a ship bringing arms for the Continental Army, Jones found she was being pursued by *Cerebus*. He deftly drew the Royal Navy away and then made his escape so that Washington could be resupplied.[91] Jones was promoted to captain and put in command of *Alfred* with tactical control over *Providence* commanded by Captain Hoysted Hacker[92].

Jones took command of *Ranger*-26 on November 1, 1777. She was pierced for eighteen on her gun deck and eight above. Jones reconfigured the ship for eighteen six pounders, a lighter approach, and sailed from Portsmouth, New Hampshire for France.[93] Jones took two prizes during the trans-Atlantic journey. Most significant, however, is that upon arriving in France, Jones and the Rangers garnered a salute from a French naval vessel, the first time a US naval vessel was so recognized. Indeed, it was the first time the Star-Spangled Banner was recognized formally by a foreign power.[94] This meant that in the eyes of the international community, the US revolution was a legitimate resistance against authoritarian power, and American sailors, whether conducting *guerre de course* or *guerre de escadre,* could not be tried as pirates.[95]

Not one to rest on his laurels, Jones got *Ranger* underway again and took two more prizes on April 14th. Next, Jones followed a practice that many insurgents employ. He leveraged his knowledge of the local area. Jones sailed for his old stomping ground of White Haven, intending to conduct a raid on the fort and set fire to the port and the vessels moored there. The raiding party departed *Ranger* in one (or more) of her small boats and landed under the cover of darkness. The group divided into two sections and spiked the guns defending the harbor. Next, as the Americans the first ship alight, several people arrived to thwart them. One of the crew decided to "jump ship" and alerted the town. The remaining rangers escaped after only torching one vessel.

Jones sailed further north to St. Mary's Isle, another environ known to him, and led a party to the estate of the Earl of Selkirk. His father worked on the grounds as a gardener (perhaps Jones would have followed in his father's footsteps had he not chosen a life at sea). He intended to capture the earl and turn him over to Continental authorities to hold for ransom of American sailors in British prisons. The earl was not at home. Hoping to eek some victory out of the day, Jones' crew demanded loot that they could claim as a prize for the venture. Against his wishes, they absconded with the estates' silver. Embarrassed and hoping to keep his honor clean, Jones purchased the lot from his crew and returned it to the Earl and Lady Selkirk, though it took him ten years.

Neither of these amphibious raids can be described as a tactical or operational success. Strategically, they were impactful. They underscored the fear that American raiders might bring the war's violence to England's shores, and they forced the admiralty to consider defensive measures.

Jones took *Drake* as he returned to France, but then ended up remaining ashore for over a year. It was not until January 1779 that he sailed in command of an old East Indies ship, converted to a war machine and renamed *Bonhomme Richard*, in honor of America's most influential representative in France, Benjamin Franklin and his work *Poor Richard's Almanac*, which was published in France as *Les Maximes du Bonhomme Richard*. From *Richards'* quarterdeck, Jones led a combined squadron of French and American sailors as they circumnavigated the British Isles. As they headed south, nearly complete with their journey, Jones saw a British merchant fleet near Flamborough Head on September 23, 1779. The fleet was escorted by *Serapis*. Realizing that he was outgunned, Jones, like a prizefighter, closed the distance to grapple and board the Royal Navy vessel, while reducing the effectiveness of her guns. *Richard* suffered fatal damage from the *Serapis* broadsides because her timbers were rotten through.[96] In naval legend, one of the crew attempted to strike the colors, only to be felled by Jones' boarding pistols. Captain Richard Pearson, then asking if Jones had lowered his colors, was met with the reply, "I have not yet begun to fight," though some historians suggest this is apocryphal. What is not in doubt is that a gunner high in the rigging onboard *Richard* dropped a grenade into *Serapis'* hold, causing a massive explosion.[97] Demoralized, the Royal Navy surrendered. *Richard* was a loss, so Jones took command of *Serapis* and, in her, returned to France.

French Naval Intervention

Several elements lent themselves to success on the high seas. First, the US coastline covers thousands of miles and twenty degree latitude. It includes several ports and innumerable navigable waterways that would be nearly impossible for the entire Royal Navy to cover or control. The nation enjoyed a significant maritime tradition, albeit one that focused more on commerce (to include smuggling as described previously) than the conduct of war. One might debate whether training an army to meet the British on the field was more daunting than converting merchantmen to men of war. If success on the high seas was an essential line of operation to assail British will and lead to their eventual withdrawal, the support of other maritime nations was an imperative to ensure this success. Benjamin Franklin sailed for France in 1776, following Silas Deane who was already in Paris acting as a secret envoy to seek their support. At first, French support was surreptitious. They allowed American ships to enter French ports to refit or enter prize claims. Funds and material needed to conduct war were sent across the Atlantic. After the Major General Horatio Gates defeated General Burgoyne at Saratoga in October 1777, the French became more confident in the colonies ability to actually defeat the British. On February 6, 1778, they signed a treaty of commerce and alliance with Benjamin Franklin.[98, 99] This expanded the American Revolution from a colonial insurgency to a world war. Before the conflict was over, not only would the French Navy conduct joint and combined operations with the Continental Army and Navy, they would join with the Spanish in fighting the Royal Navy in the Atlantic, the West Indies, and the Mediterranean.

France dispatched Charles Hector, comte d'Estaing, a general turned admiral with a squadron. It took them more than a month for them to reach the US coastline, but they arrived on March 16, 1778. The tide of the insurgency at sea rose to now include strong powers that could conduct *guerre de escadre*. It would continue to ebb and flow, with the conflict shifting between the West Indies and then back to the colonies and growing from smuggling and privateering in European waters to full on fleet engagements.

The Battle of Ushant

While d'Estaing sailed for North America, the British and French fleets would become engaged in *guerre d'escadre* in the Atlantic. The two navies had different tactics. The Royal Navy's gunnery focused on penetrating an enemy's hull to cause flooding and eventual sinking of the vessel. The French aimed at a sailing vessel's weak points, their rigging, masts, and sails. Such damage would render them unable to maneuver and thus vulnerable to continued, withering fire and eventual boarding, or in the event of a formidable foe, enabling the French vessels to withdraw.[100]

On July 27, 1778, the Royal Navy's Channel Fleet let by Admiral Keppel met with the French Brest Fleet led by Louis Guillouet, comte d'Orvilliers in the mouth of the southern end of the English Channel, one hundred miles west of the French island of Ushant. The subsequent engagement garnered the island's name and is now known as the Battle of Ushant.

When the fleets encountered one another, Keppel's thirty ships were disorganized. An element led by Admiral Sir Hugh Palliser had fallen behind the rest of the fleet by over three miles.[101] Additionally, both fleets had to deal with a storm that blew through the region. Keppel, worried that he would miss an opportunity to engage, did so aggressively. He signaled the Channel Fleet to attack the French but in a wild melee without creating an organized battle formation. Conversely, d'Orvilliers signaled his twenty-nine ships into an orderly defense, followed by effective fire that damaged the rigging and tackle of several of the Royal Navy vessels. The French suffered more casualties, seven hundred to the British five hundred, but simply the ability to engage the Royal Navy and fight another day served as a moral victory.[102] Keppel was court martialed by a kangaroo court. Fellow naval officers noted that he adhered to the permanent fighting instructions while conducting his campaign, yet was admonished anyway. There was new recognition that a naval leader may be called to task for his actions and results and could not be protected though following regulations.[103] It was another erosion of British will; a strong power's military forces damaged by morale more than gunfire.

D'Estaing's Deployment

Arriving in North America, d'Estaing encountered the British Fleet near Sandy Hook, New Jersey, the waters near the sea entrance to New York. Unfamiliar with the conditions, especially the depth or locations of shoals, he did not engage. Still, he now had valuable information. By knowing where the Royal Navy fleet was, he also knew where they were not.[104] D'Estaing sailed for Newport, Rhode Island, where only a small garrison secured the town. They surrendered merely on sighting the French. Next, d'Estaing got underway to meet Howe but was unable to engage before a storm interrupted his pursuit. He then made his way into Boston. After several weeks in North America, the French did not seem to be of much use to the Continental forces. With cold winter weather making conditions at sea and ashore more hostile, both the French and British forces sailed for warmer climate in the West Indies.[105]

The Caribbean provided friendlier weather, but the environment was still challenging. Many European immigrants suffered illness in the region; on average, a third would die within three years.[106] As an important region for trade for France and England, it became increasingly important in the conflict, and with the impending arrival of the French fleet, marked the conflict continuing to grow from a Patriot insurgency to a world war. Prior to d'Estaing and his fleet arriving British and French forces were engaged in battle in the region. On September 7, 1778, the French captured the British island of Dominica, north of their island of Martinique. The British responded in kind, invading St. Lucia, the French island south of Martinique on December 13-14. D'Estaing arrived near St. Lucia on December 14, 1778. He found the British fleet preparing a defense by anchoring their ships in a small bay called the cul-de-sac. However, to approach the waiting fleet proved too difficult. Currents between the island and Martinique to the north made maneuvering very difficult. When the French were able to engage, their fire was ineffective due to their inability to maneuver effectively.[107] D'Estaing directed his forces in an amphibious landing, but they were repelled by the British defending from fortified positions.[108] The remaining French garrison on St. Lucia was forced to surrender. In 1779, d'Estaing received additional ships and continued his campaign against British possessions and international trade, capturing St. Vincent and Grenada. An attempt was made on Barbados, but like his advance on St. Lucia, the winds and currents were not conducive to an assault. Now three British islands lay in French hands, one French island was occupied by the British. The Patriot insurgency expanded into a global war measured now not only by the powers involved but the regions to which the conflict was enjoined.

On September 9, 1779, d'Estaing with a fleet of thirty-three ships, over half of which were ships of the line anchored in the mouth of the Savannah River. They conducted an amphibious landing on Tybee Island and then laid siege to Savannah. They joined with American forces to attack the city on October 9th but were thwarted. D'Estaing and his fleet withdrew.[109] He returned to France.

Yorktown

Just as the epicenter of the naval war shifted back and forth from the colonies to the West Indies, the land campaign shifted from New England, to New York, to the south as the British sought to gain a decisive advantage that would end the conflict. At the end of December 1779, Generals Clinton and Cornwallis decided to invade the south. They departed New York and landed near Charleston with eight thousand men and laid siege to the city. It fell in May 1780. General Clinton returned to New York leaving Cornwallis to continue the southern campaign.

The French presence in North America continued to grow. The Marquis de Lafayette asked for additional troops to support the Continental Army. France sent General Rochambeau who landed in Newport, Rhode Island in July 1780 with 5,500 troops, seven ships of the line, and three frigates.[110] Washington met with Rochambeau in Hartford. He continued to emphasize two strategic concepts:

> 1st, That there can be no decisive enterprise against the maritime establishments of the English in this country without a constant naval superiority; and 2nd, That all of the enterprises which may be undertaken, the most important and decisive is the reductions of New York, which is the center focus of all the British forces.

Washington's recognition that naval superiority was an imperative would continue to prove to be correct. Fortunately, as the Continental Navy declined, the French Navy increased its presence. In March 1781, Rear Admiral Francois Joseph Paul, Comte de Grasse, led a large fleet from Brest. Cornwallis had marched his men through the Carolinas into Virginia and established a position in Yorktown that would enable resupply through the Chesapeake Bay. Rochambeau wrote to Comte de Grasse and suggested that he attack there. De Grasse subsequently sent word to Washington that was received on August 14, 1781 informing the general that de Grasse had three thousand troops and twenty-nine ships of the line, and he intended to sail to the Chesapeake.[111] De Grasse arrived at the mouth of the Chesapeake on August 30, landing troops on the James River. On September 5, a fleet was spotted at sea.

Surmising this was the Royal Navy, de Grasse got underway to engage them. His leading opponent was Admiral Thomas Graves. As the two elements met, they formed a large "v" rather than a traditional set of parallel lines for a fleet engagement. Ships that were close together endured a more vicious battle than those at the rear of the formation that had more distance between opponents. The Royal Navy endured heavier damage and casualties and withdrew. De Grasse sailed back into the Chesapeake, cutting off Cornwallis ability to resupply or withdraw from the sea.

Washington and Rochambeau combined forces and led an army of six thousand men that arrived in Williamsburg, Virginia on September 14, 1781 and rendezvoused with another force led by Lafayette. Cornwallis and his forced were placed under siege and capitulated on October 19, 1781 after a final attack by the combined forces.

While the global war would continue until the signing of the Treaty of Paris in 1783, Cornwallis' surrender marked the end of the war in the colonies. The Battle of Yorktown serves as a metaphor for the entire conflict. As Washington suggested, victory over the British could not succeed without superiority at sea; similarly, the Patriot insurgency could not be successful without the assistance of strong foreign powers to initially supply and finally fight alongside the Continentals.

ENDNOTES

1. Eric Robson, *The American Revolution in its Political and Military Aspects, 1763-1783* (New York: Norton & Co., 1966), 3.
2. Rober Middlekauff, *The Glorious Cause: the American Revolution, 1763-1789* (Oxford: Oxford University Press, 2007), 70.
3. Ibid.
4. Ibid., 74-75.
5. Alan Taylor, *American Revolutions: A Continental History, 1750-1804* (New York: W. W. Norton & Co., 2016), 96-97.
6. These particular mobs acted on their own without being directed by the Otis faction. After this destruction, Samuel Adams worked to bring the mobs more securely under his management.
7. Edmund S. Morgan and Helen M. Morgan, *The Stamp Act Crisis: Prologue to Revolution* (University of North Carolina Press, 1995), 125-149.
8. In New Providence, Connecticut, for example, a mob forced a stamp distributor into a coffin, lowered it into a grave, and began to shovel dirt onto it until the man cried out and agreed to resign his commission.
9. Middlekauff, *The Glorious Cause: the American Revolution, 1763-1789*, 105-106.
10. John Dickinson, *Letters from a Farmer in Pennsylvania* [Kindle] (2013).
11. The name "Sons of Liberty" was bestowed on various groups of Patriots during and after the Stamp Act crisis. The Sons in each colony sought to communicate and coordinate their actions with each other. In modern parlance they served as the insurgent underground, particularly in urban areas.
12. Middlekauff, *The Glorious Cause: the American Revolution, 1763-1789*, 174-185.
13. For detailed and incisive analysis of the massacre, see Erich Hinderaker, Boston's Massacre (Harvard University Press, 2017). The author discusses the political implications as well as analyzing the confusing firsthand accounts of the conflict. Hiller B. Zobel's The Boston Massacre examines the events that predated the incident to show how political passions formed and sharpened in the days leading up to it.
14. Patronage to officers was one of several motivations. Most of the troops were needed in Canada and around the Great Lakes. In 1768, Boston became the single source of a significant number of soldiers. The numbers of British soldiers in New York was comparatively small.
15. Hiller B. Zobel, *The Boston Massacre* (New York: W. W. Norton & Co., 1996), 132-44.
16. For a thorough treatment of the personalities and political dynamics behind the Tea Party, see Harlow Giles Unger, *American Tempest: How the Boston Tea Party Sparked a Revolution* (Da Capo Press, 2011).
17. Middlekauff, *The Glorious Cause: the American Revolution, 1763-1789*, 229-231.
18. Ibid., 200.
19. Carol Sue Humphrey, "Top 10 Revolutionary Newspapers," *Journal of the American Revolution* (26 February 2015), para. 3. https://allthingsliberty.com/2015/02/top-10-revolutionary-war-newspapers/.
20. Episcopal government is rule by bishops, such as the Anglican Church in England practiced. The Scottish Puritans pursued a Presbyterian system in which selected elders governed the church.

[21] Ernest Lee Tuveson, *Redeemer Nation: the Idea of America's Millennial Role* (Chicago: University of Chicago Press, 1968), 52-90.

[22] Middlekauff, *The Glorious Cause: the American Revolution, 1763-1789*, 4.

[23] William T. Youngs, *The Congregationalists. Denominations in America* (Westport, Connecticut: Praeger, 1998), 114.

[24] Middlekauff, *The Glorious Cause: the American Revolution, 1763-1789*, 111.

[25] Taylor, *American Revolutions: a Continental History, 1750-1802*, 5.

[26] Henry St. John Bolingbroke, *Letters on the Spirit of Patriotism and on the Idea of a Patriot King* (Oxford, UK: Clarendon Press, 1926).

[27] Ibid., 3-4.

[28] T. H. Breen, *American Insurgents, American Patriots: The Revolution of the People* (New York: Hill and Wang, 2010), 241.

[29] John Fiske. *The American Revolution: In Two Volumes*, Houghton, Mifflin and Company, 1892, p. 323.

[30] Glenn Grasso, "Creating the Continental Navy," C-SPAN video, 50:41, November 26, 2012, https://www.c-span.org/video/?309570-1/creating-continental-navy.

[31] Ibid.

[32] Allan Westcott, ed., *American Sea Power since 1775* (New York: J.B. Lippincott Company, 1947), 4.

[33] Westcott, *American Sea Power since 1775*, 7.

[34] Robert H. Patton, *Patriot Pirates: The Privateer War for Freedom and Fortune in the American Revolution* (New York: Pantheon Books, 2008.), 42.

[35] Patton, *Patriot Pirates*, 34.

[36] Privateers and Mariners in the Revolutionary War" < http://www.usmm.org/revolution.html>

[37] Westcott, *American Sea Power since 1775*, 8.

[38] Alfred Thayer Mahan, *Major Operations of the Navies of the War of American Independence* (New York: Greenwood Press Publishers, 1969), 60.

[39] Edgar Stanton Maclay, *A History of American Privateers* (New York: D. Appleton and Company, 1899), x-xi.

[40] Sam Willis, *The Struggle for Sea Power: A Naval History of the American Revolution* (New York: WW. Norton and Company, 2015), 29.

[41] Ibid., 29.

[42] Sheldon Whitehouse, "1772 Gaspee Affair," CSPAN video, 13:32, June 9, 2015, https://www.c-span.org/video/?326431-3/discussion-1772-gaspee-affair.

[43] Ibid.

[44] Ibid.

[45] McGrath, *Give Me A Fast Ship: The Continental Navy and America's Revolution at Sea* (New York: Berkley Caliber, 2014), 16.

[46] Willis, *The Struggle for Sea Power*, 58.

[47] Ibid., 59.

[48] William Fowler Jr., *Rebels Under Sail: The American Navy during the Revolution* (New York: Charles Scribner's Sons, 1976), 156-57.

[49] Westcott, *American Sea Power since 1775*, 5.

50 Ibid., 5.

51 Maclay, *A History of the United States Navy*, 49-50.

52 Willis, *The Struggle for Sea Power*, 99.

53 Ibid., 80-81.

54 McGrath, *Give Me Fast Ship*, 18-19.

55 Ibid., 58.

56 Tim McGrath, *John Barry: An American Hero in the Age of Sail* (Yardley: Westholme Publishing, 2010), 59.

57 Westcott, *American Sea Power since 1775*, 5-6.

58 Ibid., 6.

59 Willis, *The Struggle for Sea Power*, 88.

60 Joel W. Thurman, "The Revolution Told by One of the Navy's Greatest Ships," *Journal of the American Revolution* (16 November 2016). https://allthingsliberty.com/2016/11/revolution-told-one-navys-greatest-ships/.

61 Maclay, *A History of the United States Navy*, 39.

62 Westcott, *American Sea Power since 1775*, 6.

63 Maclay, *A History of the United States Navy*, 40.

64 Court martials were common in the age of sail, serving first as a board of inquiry to gather the facts relaying to a deployment. Officers often requested court martials be formed to examine their own actions.

65 Willis, 47; Orlando W. Stephenson, "The Supply of Gunpowder in 1776," *American Historical Review* 30, no 2. (1925): 273.

66 Willis, *The Struggle for Sea Power*, 43.

67 Ibid., 43-44; David Syrett, *The Royal Navy in American Waters 1775-1783* (Aldershot: Scolar Press, 1989), 22; Andrew Jackson O'Shaughnessy, *The Men Who Lost America: British Leadership, the American Revolution, and the Fate of the Empire* (New Haven: Yale University Press, 2013), 14.

68 Willis, *The Struggle for Sea Power*, 45.

69 Ibid., 70.

70 Westcott, *American Sea Power since 1775*, 10.

71 Ibid., 12.

72 Willis, *The Struggle for Sea Power*, 152.

73 Westcott, *American Sea Power since 1775*, 12.

74 Willis, *The Struggle for Sea Power*, 146.

75 E.B. Potter, ed., *Sea Power: A Naval History (Second Edition)* (Annapolis: Naval Institute Press, 1981), 35; Westcott, *American Sea Power since 1775*, 12.

76 Westcott, *American Sea Power since 1775*, 12.

77 Ibid., 15.

78 Ibid., 15.

79 Patton, *Patriot Pirates*, 172.

80 Ibid., 173.

81 Ibid., 175.

82 Ibid., 174.

83 Ibid., 176.

84 Ibid., 178-179.

85 Ibid., 183.

86 Ibid., 196.

87 Ibid., 199.

88 During the age of sail, physical punishments such as whipping the guilty with a lash or 'cat of nine tails' occurred.

89 "First Lieutenant" in naval parlance is not a rank, but a role onboard ship, the officer in charge of the deck division(s), the sailors that operate and maintain the ships main deck equipment, lines, and small boats. Naval ships in the age of sail when first mentioned are written with the number of guns included, thus Alfred had 24 guns.

90 James Fenimore Cooper and Dennis M. Conrad, *John Paul Jones: Three Histories* (Sarajevo: Seea Publishing, 2013).

91 Maclay, *A History of the United States Navy*, 46.

92 All commanding officers in naval tradition, no matter the rank are referred to as "captain." It is not clear if Hacker was also promoted to captain, as Jones had been, or if he was perhaps, a lieutenant in command of Providence.

93 Cooper and Conrad, *John Paul Jones*.

94 In the age of sail, sailors were often referred to by the name of their ship, hence "Rangers" for the crew of the Ranger or "Alfreds" for the crew of Alfred.

95 Cooper and Conrad, *John Paul Jones*.

96 Westcott, *American Sea Power since 1775*, 20.

97 Ibid., 21.

98 Potter marks the date at July 23, 1778. Most sources state the battle occurred on July 27.

99 Potter, *Sea Power*, 37.

100 Ibid., 41.

101 Andrew Lambert, *War at Sea in the Age of Sail* (London: Cassell, 2000), 133.

102 Potter, *Sea Power*, 41; Lambert, *War at Sea in the Age of Sail*, 133.

103 Potter, *Sea Power*, 41.

104 Willis, *The Struggle for Sea Power*, 240.

105 Ibid., 248.

106 Ibid., 254.

107 Ibid., 262.

108 Ibid., 262.

109 William Laird Clowes, *The Royal Navy: A History from the Earliest Times to the Present* (London: Sampson Low, Martson and Company, 1899), 32.

110 Potter, *Sea Power*, 43.

111 Ibid., 45

CHAPTER 7.
GOVERNMENT COUNTERMEASURES

INITIAL RESPONSE TO INSURGENCY

The actions of the British Parliament in the years up to the violent upheaval of 1775-1776 fall into two categories differentiated by intent. First, there were a series of legislative acts designed to provide budgetary relief in answer to the massive debt that Great Britain acquired in the course of the Seven Years War. The Sugar Act, Currency Act, and Stamp Act were each aimed at improving the government's revenues and were not designed as punitive. Following the Stamp Act crisis, however, Parliament began to craft legislation intended to communicate to the people of colonial North America that Great Britain was their master and must be obeyed.

The first of these punitive acts was the Declaratory Act in 1766—a move that in retrospect appears unwise. The sudden, sustained colonial rebellion had taken British gentry by surprise. The fundamental illogic of empire had come home to roost. Great Britain's success in creating a global network of trade had resulted in burgeoning population along the Atlantic seaboard—but one that was not content with the feudal relationship with the mother country that Parliament had in mind. The American colonists were in every sense a modern society brought politically alive by Enlightenment philosophy. When the gentlemen of Parliament first came into contact with the colonists' consequent political vigor, they responded with the hamhanded Declaratory Act. It had all the sophistication of wishful thinking and was wholly ineffective in solving the growing problem.

Instead, London discovered that the colonists were capable of a surprising capacity for unified action. Colonial boycotts, non-importation resolutions, and determination to extend smuggling operations in defiance of the law frustrated government ministers, who seemed capable of only one response: coercion. The resulting acts of Parliament were a combination of attempts to rescue the government's budget and punish the colonial upstarts. The Townshend Acts, in turn, led to a deeper crisis.

In the years leading up to the Revolution, both sides likewise attempted to use economic suasion to influence the other. Great Britain was predisposed to lose these contests because of the asymmetric economic situations in England and America. The British people—and especially the political elites—were vulnerable to any disruption of the imperial economy. The gentlemen of Parliament were mostly monied gentlemen with far-flung interests in foreign markets. Colonists craved luxury goods from England and Europe, and they needed foreign markets for their agricultural goods and raw resources. Nevertheless, Americans—particularly if they could get supplies from the French—were were able to survive without British commerce when they had to. Their embryonic manufacturing capability would only grow under the pressure of non-importation and war. In the meantime, many chose to wear rough colonial homespun in place of the fancy silks and linens from London.

FROM POLITICAL TO MILITARY RESPONSE

The Townshend Acts predictably resulted in another round of colonial acts of rebellion. Blinded by their cultural prejudice, Parliament replied to popular resistance with the decision to station British regulars in Boston. Ministers could console themselves by proposing that the troops were there for the colonists' protection and that, therefore, the colonists should foot the bill. However, the message on the other end of the transaction was clear: Americans are inferiors who will have to be roughed up to behave themselves. The insult was an outrage to the citizens of Boston.

British regulars were products of eighteenth-century military training, which honed them for battle on conventional battlefields. Although some had developed skills in light infantry units—particularly during the Seven Years War in America—the dominant training emphasized how to march, load and fire their muskets, and keep their kit clean. They did not teach how to deal with riots and demonstrations, or how to win hearts and minds. Sent to keep law and order and to obey the directions of the royal governor, the British troops aggravated the situation by allowing trouble-making citizens to goad them into violence. The resulting Boston Massacre of 1770 was a perfect propaganda tool for the Patriots.

Throughout the Revolutionary War, Parliament and her generals clung to the notion that British troops in America's major cities, backed up by Royal Navy warships in her harbors, would quell the violence and restore royal authority. When this proved not to be the case, a succession of generals—Gage, Howe, Burgoyne, Clinton, Cornwallis, and others—struggled to find the military solution to a political problem. They fixated on repeated attempts to destroy the Continental Army, hoping that a decisive battlefield defeat would translate into a political surrender. This was almost certainly untrue, but in any case, that defeat never came. The Patriots suffered some military setbacks—Long Island, Brandwine, Germantown—but they also inflicted impressive defeats on the British at Trenton, Princeton, Bemis Heights, and Yorktown.

CHAPTER 8.
CONCLUSION

TRANSITION FROM INSURGENCY TO GOVERNANCE

One of the most difficult tests for an insurgency is the question as to whether it can transform from an illegal, clandestine network into a legitimate government. The transition is difficult because the skills and experience required are different. The Patriot insurgency had advantages that set them up for success at the conclusion of the Revolutionary War. First, in nearly every case, the primary Patriot leaders already had years of experience in governing. Most of them were involved in colonial legislatures prior to the outbreak of hostilities. Second, the crisis itself served as a catalyst to force the Patriots into the role of a shadow government before the start of the war. The Intolerable Acts of 1774 led to the dissolution of local assemblies and the consequent operation of a host of illegal and extra-legal activities related to governance. Finally, the Continental Congress, by the end of the war, had been in operation and functioning as a national government for eight years.

The real question for the new country was not whether they could govern themselves, but *how* they intended to govern and to what degree the United States would be united. In July 1776, concurrent with declaring independence, the Continental Congress drafted the Articles of Confederation. This pragmatic, initial attempt at achieving a national government barely sufficed to prosecute the war, and the American military suffered throughout the conflict from lack of funds and logistical support. The Congress was able to conclude the strategically important French alliance, but the conclusion of the conflict left in doubt the viability of the new nation to navigate the tricky and dangerous waters of international relations.

The inadequacy of the Articles—in particular the government's lack of a powerful executive and its inability to tax—was obvious to all. What was not so clear was the solution. The Patriots, after all, had just spilled much blood to free themselves from the tyranny of a government that was so centralized—and seemingly so corrupt—that it could not (or would not) respond to the legitimate needs of the governed. Indeed, it was not so much the United States that rebelled but thirteen semi-united colonies. Most citizens of the fledgling union reserved their primary loyalty for their state, not to any supposed American nation. Sectional differences—and the slavery issue in particular—underscored the differences that divided the states. To put the country's future into the hands of a remote, central government might nullify the very gains that the war had won.

Still, the common danger felt by all thirteen states had the potential to unify them. There was no telling what a reinvigorated England might do once it extricated itself from international war and recovered. Nor were the intentions of France, Spain, and even Russia yet clear regarding who would control the New World. The European powers had not relinquished their claims on Canada, the Mississippi, New Orleans, or the West. No matter which of them might assert themselves in these key regions, a European interest there would endanger the colonists' future expansion. Worse, distant powers could pursue intrigues with the Indians or even pit one state against another. It would be impossible for each of the thirteen newly independent states to conclude meaningful treaties. Instead, one unified national government would have to do that, with the power to negotiate, ratify, and enforce treaties.

Likewise, there was the massive problem of war debt. The several states had, for lack of a powerful national government, funded the war at the state level through bonds, loans, and deficit spending. State debts ran the gamut from manageable to crippling, and all saw the advantages of handing the problem over to a national government that could both tax and secure advantageous loans from overseas. The proposed new constitution would have as one of its key selling points, the national government's assumption of the states' debts.

The delegates to the 1787 Constitutional Convention met in closed session so as to ease the path forward through the sticky political obstacles to a new constitution. Opportunities for failure abounded. How were the delegates to merge thirteen very different states together when some, like Rhode Island and Delaware, were tiny in both size and population, and others, primarily Virginia, were huge? If power were apportioned according to population, the small states would have no say in the country's course. If each state had equal representation, the small states would have disproportionate influence. Slavery complicated matters even more because the southern states reckoned themselves dependent upon not only the continued existence of the institution but also the expansion of it. Others saw the inherent illogic in founding a republic on the purest principles of liberty and the natural civil rights that everyone was supposed to enjoy and yet, on the other hand, keeping alive the slave trade that European powers had abandoned.

The solution to these and many other hard issues was twofold. The explicit part of the solution was compromise. The tacit part was postponement. Compromise would give rise to a bicameral national legislature with the upper house favoring smaller states and the lower house favoring larger states. The potentially deadly issue of slavery was likewise given over to compromise, but resolution of the fundamental divide between the planter elite of the south and the more commercialized north was merely delayed. The Three-Fifths Compromise (which counted slaves as three-fifths of a person for the purpose of representation in Congress) and the decision to balance newly acquired states between slave and non-slave was a temporary measure designed to hold off the question until the country could get on its feet. The civil war that followed a little over seventy years later revisited the explosive issue and resolved it in the blood of 620,000 war dead. Only after the political and social turmoil of presidential and congressional reconstruction was the issue of governance resolved.

THE UNSEEN SEED OF REBELLION

Beyond a detailed analysis of the causes of the Patriot insurgency and the revolution that followed, there is one other factor that histories largely pass over. One of the main causes of the American Revolution was success—specifically, Britain's success as a world empire and America's success as a confederation of thriving colonies.

It is possible to imagine different scenarios than that which occurred. If Great Britain had lost the Seven Years' War, and France had strengthened its grip on Canada, the Mississippi, and the Caribbean, it would have presented an existential challenge both to England and her colonies in North America. The threat of Anglicanism would have paled against that of hostile French Roman Catholics putting a stranglehold on the thirteen colonies and threatening both their trade and their desired expansion westward. Facing a common enemy, the colonies and Parliament would likely have forged an alliance aimed at survival.

Likewise, the economic and political success of the colonies themselves fed the impulse toward independence. If we instead imagine that the economic depression after the Seven Years' War deepened and permanently scarred trade and growth, it would have sapped the vigor that ultimately turned against the mother country. The colonies would have remained economically dependent on England's protection and markets, and a break with the empire would have been inconceivable. Alternately, it is a simple matter to imagine a scenario in which sectional, confessional, and class distinctions overcame any desire among the colonies to unite. A politically fragmented set of colonies could have been manipulated by Parliament and pitted one against another in the struggle.

Instead, it was Britain's global success and the colonies economic and political health that led to revolution. The logic of the British Empire had run its course, with the result that Parliament was attempting to rule a far-flung enterprise using home-court rules. There was a conspicuous lack of reflection among members of Parliament as to the dangers of success and expansion. There seemed to be no discussion of how the House of Commons might reform itself into an imperial legislature that could respond to empire-wide issues instead of to petty factions in London. Across the Atlantic, Maslow's hierarchy of needs played out as Americans rose up from subsistence farming and a struggle to survive to become an economic powerhouse. Freed from the needs for shelter, food, and security (once France had been defeated and the Indians chased westward), Americans discovered political sensitivities that might otherwise have never appeared.

Success demanded a change in perspective that led ultimately to revolution, war, and independence. This phenomenon is a key factor in understanding the Patriot insurgency, as well as others in history. Insurgency does not emerge solely from shared deprivation. It also grows from shared success.

BIBLIOGRAPHY

BIBLIOGRAPHY

Anderson, Benedict. *Imagined Communities*. London: Verso, 2006.

Anderson, Fred. *Crucible of War: The Seven Years' War and the Fate of Empire in British North America, 1754 – 1766*. New York, Vintage Books: 2001.

Bos, Nathan. "Underlying Causes of Violence." In *Human Factors Considerations of Undergrounds in Insurgencies*, edited by Nathan Bos. Fort Bragg, NC: United States Army Special Operations Command, 2013.

Breen, T. H. *American Insurgents, American Patriots: The Revolution of the People*. New York: Hill and Wang, 2010.

Buhaug, Halvard and Jan Ketil Rød. "Local Determinants of African Civil Wars, 1970–2001." *Political Geography* 25, no. 3 (2006): 315-35.

Calloway, Colin G. *The American Revolution in Indian Country: Crisis and Diversity in Native American Communities*. Cambridge, Cambridge University Press, 1995.

— *Crown and Calumet: British-Indian Relations, 1783-1815*. Norman, OK: University of Oklahoma Press, 1987.

— "Suspicion and Self-Interest: The British-Indian Alliance and the Peace of Paris." *The Historian* 48, no. 1 (November 1985): 41-60.

— "'We Have Always Been the Frontier': The American Revolution in Shawnee Country." *American Indian Quarterly* 16, no. 1 (Winter 1992): 39-52.

Clowes, William Laird. *The Royal Navy: A History from the Earliest Times to the Present*. London: Sampson Low, Martson, and Company, 1899.

Collier, Paul and Anke Hoeffler. "Greed and Grievance in Civil War." *Oxford Economic Papers* 56, no. 4 (2004): 563-95.

Cooper, James Fenimore and Dennis M. Conrad. *John Paul Jones: Three Histories* Sarajevo: Seea Publishing, 2013.

Countryman, Edward. "Indians, the Colonial Order, and the Social Significance of the American Revolution." *The William and Mary Quarterly*. 53, no. 2 (April 1996): 343-62.

Crossett, Chuck, ed. *Casebook on Insurgency and Revolutionary Warfare Volume II: 1962–2009*. Fort Bragg, NC: United States Army Special Operations Command, 2012.

Crownover, W. B. Max. "Complex System Contextual Framework (CSCF): A Grounded-Theory Construction for the Articulation of System Context in Addressing Complex Systems Problems." [PhD Dissertation]. Norfolk, VA: Old Dominion University, 2005.

Diani, Mario. "The Concept of Social Movement." *Sociological Review* 40, no. 1 (1992): 1–25.

Diani, Mario and Ivano Bison. "Organizations, Coalitions, and Movements." *Theory and Society* 33 (2004): 281–309.

Dickinson, John. *Letters from a Farmer in Pennsylvania*. [Kindle] 2013.

Dowd, Gregory Evans. *A Spirited Resistance: The North American Indian Struggle for Unity, 1745–1815*. Baltimore: Johns Hopkins University Press, 1992.

Draper, Theodore. *A Struggle For Power: The American Revolution*. New York: Vintage, 1997.

Fearon, James D. and David D. Laitin. "Ethnicity, Insurgency, and Civil War." *American Political Science Review* 97, no. 1 (2003): 75–90.

Fitz, Caitlin A. "'Suspected on Both Sides': Little Abraham, Iroquois Neutrality, and the American Revolution," *Journal of the Early Republic* 27, no. 3 (Fall 2008): 299-335.

Fowler, William. *Rebels Under Sail: The American Navy during the Revolution*. New York: Charles Scribner's Sons, 1976.

Goldstone, Jack A., Robert H. Bates, David L. Epstein, Ted Robert Gurr, Michael B. Lustik, Monty G. Marshall, Jay Ulfelder, and Mark Woodward. "A Global Model for Forecasting Political Instability." *American Journal of Political Science* 54, no. 1 (2010):190–208.

Grasso, Glenn. "Creating the Continental Navy," C-SPAN video, 50:41. Aired 26 November 2012, https://www.c-span.org/video/?309570-1/creating-continental-navy.

Graymont, Barbara. *The Iroquois in the American Revolution*. Syracuse, NY: Syracuse University Press, 1972.

Gurr, Ted Robert. *Why Men Rebel*. Princeton, NJ: Princeton University Press, 1970.

Hinderaker, Eric. *Elusive Empires: Constructing Colonialism in the Ohio Valley, 1673 – 1800*. Cambridge, MA: Cambridge University Press, 1997.

Humphrey, Carol Sue. "Top 10 Revolutionary Newspapers." *Journal of the American Revolution* (26 February 2015), para. 3. https://allthingsliberty.com/2015/02/top-10-revolutionary-war-newspapers/.

Huntington, Samuel P. *Political Order in Changing Societies.* New Haven: Yale University Press, 1968.

Joshi, Madhav and David Mason. "Between Democracy and Revolution: Peasant Support for Insurgency versus Democracy in Nepal." *Journal of Peace Research* 45, no. 6 (2008): 765–782.

Kalyvas, Stathis N. *The Logic of Violence in Civil War.* New York: Cambridge University Press, 2006.

Lambert, Andrew. *War at Sea in the Age of Sail.* London: Cassell, 2000.

Levinson, David. "An Explanation for the Oneida-Colonist Alliance in the American Revolution." *Ethnohistory* 23, no. 3 (Summer 1976): 265-89.

Lichbach, Mark. *The Rebel's Dilemma.* Ann Arbor: University of Michigan Press, 1995.

Maclay, Edgar Stanton. *A History of the United States Navy.* New York: D. Appleton and Company, 1898.

— *A History of American Privateers.* New York: D. Appleton and Company, 1899.

Mahan, Alfred Thayer. *Major Operations of the Navies of the War of American Independence.* New York: Greenwood Press Publishers, 1969.

Manogaran, Chelvadurai and Bryan Pfaffenberger, "Introduction: The Sri Lankan Tamils." In *The Sri Lankan Tamils: Ethnicity and Identity*, Boulder, CO: Westview Press, 1994.

Mason, David. "Land Reform and the Breakdown of Clientelist Politics in El Salvador." *Comparative Political Studies* 8, no. 4 (1986): 487–517.

McAdam, Doug, Sidney G. Tarrow, and Charles Tilly. *The Dynamics of Contention.* Cambridge, UK: Cambridge University Press, 2001.

McGrath, Tim. *Give Me A Fast Ship: The Continental Navy and America's Revolution at Sea.* New York: Berkley Caliber, 2014.

— *John Barry: An American Hero in the Age of Sail.* Yardley, PA: Westholme Publishing, 2010.

Middlekauff, Robert. *The Glorious Cause: The American Revolution, 1763-1789.* Oxford: Oxford University Press, 1982.

Morgan, Edmund S. and Helen M. Morgan. *The Stamp Act Crisis: Prologue to Revolution.* Chapel Hill, NC: University of North Carolina Press, 1995.

O'Shaughnessy, Andrew Jackson. *The Men Who Lost America: British Leadership, the American Revolution, and the Fate of the Empire.* New Haven: Yale University Press, 2013.

Paige, Jeffery. *Agrarian Revolution.* New York: The Free Press, 1975.

Parmenter, Jon William. "Pontiac's War: Forging New Links in the Anglo-Iroquois Covenant Chain, 1758-1766." *Ethnohistory* 44, no. 4 (Autumn 1997): 617-54.

Patton, Robert H. *Patriot Pirates: The Privateer War for Freedom and Fortune in the American Revolution.* New York: Pantheon Books, 2008.

Potter, E.B. ed. *Sea Power: A Naval History (Second Edition).* Annapolis, MD: Naval Institute Press, 1981.

Raleigh, Clionadh, Andrew Linke, Håvard Hegre, and Joakim Karlsen. "Introducing ACLED: An Armed Conflict Location and Event Dataset: Special Data Feature." *Journal of Peace Research* 47, no. 5 (2010): 651-60.

Robson, Eric. *The American Revolution in its Political and Military Aspects, 1763-1783.* New York: Norton & Co., 1966.

Rocca, Al M. "The Impact of Geography on the American Revolution: Expanding Regions of British Military Responsibility." *Social Studies Review* 42, no. 2 (Spring 2003). https://www.questia.com/magazine/1P3-345763931/the-impact-of-geography-on-the-american-revolution.

Sambanis, Nicholas. "Do Ethnic and Nonethnic Civil Wars Have the Same Causes?: A Theoretical and Empirical Inquiry (Part 1)." *Journal of Conflict Resolution* 45, no. 3 (2001): 259-82.

Schmidt, Ethan A. *Native Americans in the Revolution: How the War Divided, Devastated, and Transformed the Early American Indian World.* Santa Barbara, CA: ABC-CLIO, 2014.

Snapp, J. Russell. *John Stuart and the Struggle for Empire on the Southern Frontier.* Baton Rouge, LA: Louisiana State University Press, 1996.

St. John Bolingbroke, Henry. *Letters on the Spirit of Patriotism and on the Idea of a Patriot King.* Oxford, UK: Clarendon Press, 1926.

Stephenson, Orlando W. "The Supply of Gunpowder in 1776." *American Historical Review* 30, no 2. (1925): 271-81.

Syrett, David. *The Royal Navy in American Waters 1775-1783.* Aldershot, UK: Scolar Press, 1989.

Taylor, Alan. *American Colonies: The Settling of North America.* New York: Penguin Books, 2001.

— *American Revolutions: A Continental History, 1750-1804.* New York: W. W. Norton & Co., 2016.

Thurman, Joel W. "The Revolution Told by One of the Navy's Greatest Ships." *Journal of the American Revolution* [online] (16 November 2016). https://allthingsliberty.com/2016/11/revolution-told-one-navys-greatest-ships/.

Tilly, Charles. "Why and How History Matters." In *The Oxford Handbook of Contextual Political Analysis*, edited by Robert E. Gordon and Charles Tilly [online]. Oxford: Oxford University Press, 2006.

Tilly, Charles and Robert E. Gordon. "It Depends." In *The Oxford Handbook of Contextual Political Analysis*, edited by Robert E. Gordon and Charles Tilly [online]. Oxford: Oxford University Press, 2006. DOI:10.1093/oxfordhb/9780199270439.003.0001

Tilly, Charles and Sidney Tarrow. *Contentious Politics.* Boulder, CO: Paradigm Publishers, 2007.

Tiro, Karim M. "A 'Civil War? Rethinking Iroquois Participation in the American Revolution." *Explorations in Early American Culture* 4 (2000): 148-65.

Tuveson, Ernest Lee. *Redeemer Nation: the Idea of America's Millennial Role.* Chicago: University of Chicago Press, 1968.

Unger, Harlow Giles. *American Tempest: How the Boston Tea Party Sparked a Revolution.* Cambridge, MA: Da Capo Press, 2011.

Westcott, Allan, ed. *American Sea Power since 1775.* New York: J.B. Lippincott Company, 1947.

White, Richard. *The Middle Ground: Indians, Empires, and Republics in the Great Lakes Region, 1650 – 1815.* Cambridge: Cambridge University Press, 1991.

Whitehouse, Sheldon. "1772 Gaspee Affair," CSPAN video, 13:32. Aired 9 June 2015. https://www.c-span.org/video/?326431-3/discussion-1772-gaspee-affair.

Willis, Sam. *The Struggle for Sea Power: A Naval History of the American Revolution.* New York: WW. Norton and Company, 2015.

Youngs, William T. *The Congregationalists. Denominations in America.* Westport, Connecticut: Praeger, 1998.

Zobel, Hiller B. *The Boston Massacre.* New York: W. W. Norton & Co., 1996.

INDEX

A

Acadia 70
Adams, John 5, 111, 124, 133, 151
Adams, Samuel (Sam) 93, 113, 122
Afghanistan 9
African slaves 91
Agworondougwas, Peter (Good Peter) 53
Ahaya of Cuscowilla (Cowkeeper) 58
Alabama 57
Albany, New York 30
Alfred (Continental Navy) 151, 152, 158
Algonquian speakers 44, 65
Allen, Ethan 149, 153–154
Allen, William 40
American colonies 28, 37–43
    Articles of Association 17
    British occupation of 110–111, 114
    British policy toward 92–94
    class distinctions 83–85
    Committees of Correspondence 112, 118, 133
    Committees of Safety 118
    distance from England 28
    geography 120
    governance 177–179
    infrastructure 28–29
    Middle colonies 27, 29, 31
    navies 147–148
    New England colonies 27
    population 83–86
    population density 84
    Southern colonies 27, 32
    success as a confederation 179–180
    trade with sugar islands 91
    war debt 178
American Revolution
    catalyst that led to 69
    causes 179–180
    character 132–133
    as civil war 4, 132
    expansion to world war 160
    geography 27–33
    historical material 5
    land operations 5, 134–135, 144
    Native American view of 43, 51–52
    naval operations 5, 143, 144–164
    northern campaigns 137–141, 144
    physical environment 9
    shot heard 'round the world 18
    southern campaigns 141–144
    support for 133
    timeline 18, 19
    transition to governance 177–179
Amherst, Jeffrey 48, 66, 103
Anderson, Benedict 13
Andrea Doria (Continental Navy) 152
Andros, Edmund 51
Anglicans and Anglicanism 112, 124, 125, 179
    vs Congregationalists 85
    in Middle colonies 86
    Mohawk Mission 51
    in Southern colonies 86
    in Virginia 38
Anglo-Cherokee war of 1759-1761 55
Anglophiles 51
Annus Mirablis 16
anocracy 15
Appalachian Mountains 6, 16, 39, 48, 104, 131
ARIS Tier 1 Insurgency Case Studies 3
    methodology 7–15
    sections 6–7

Arnold, Benedict 135–137, 149–150, 153–154
   Battle of Valcour Island 155
Articles of Association 17
Articles of Confederation 19, 177
Atlantic Ocean 145, 160
Attakullakulla (Little Carpenter) 55
Augusta, Georgia 56
autocracy 15
auxiliary and irregular forces 5
Aztecs 46

B

Bahamas 152
Baptists 86
Barbados 162
Barry, John 151, 157
Battle of Brooklyn 137
Battle of Bunker Hill 18, 116
Battle of Fort William Henry 70
Battle of Long Island 18, 28, 138, 173
Battle of Monmouth 19, 140
Battle of Moore's Creek 38
Battle of Oriskany 53
Battle of Princeton 138
Battle of Quebec 30, 137
Battle of Quiberon Bay 70
Battle of the Saintes 19
Battle of Trenton 18, 138, 173
Battle of Ushant 160–161
Battle of Valcour Island 155
Battle of Waxhaws 141
Battle of Yorktown 143–144, 163–165, 173
Battles of Lexington and Concord 18, 116, 122–123
Bay of Biscay 155
Bemis Heights 139, 173

Berkeley, Lord of Stratton 40
Bingham, William 150
Bird, Henry 61
Blackfish 61
The Black Prince 151
black population 85
Bolingbroke, Henry 128
Bonhomme Richard (Continental Navy) 159
Boone, Daniel 61
Boston (Continental Navy) 152
Boston Chronicle 122
Boston Evening Post 122
Boston Gazette 121–122
Boston, Massachusetts 31, 41
   British occupation 31, 110–111, 123, 172
   British withdrawal 137
   "Circular Letter" 109
   port closure 114
   response to Coercive Acts 122
   Stamp Act crisis 106
   timeline 17
Boston Massacre 17, 110–111, 122, 172
Boston Non-Importation Agreement 17
Boston Tea Party 17, 111–117, 122
boycotts 114, 130, 171. *See also* nonimportation movement
Braddock, Edward 120
Bradford, Thomas 122
Bradford, William 122
Brandywine Creek 139, 173
Brant, Joseph (Thayendanegea) 51–52, 140
Brant, Molly 51
Breed's Hill (Bunker Hill) 18, 116
British Canada 71, 103, 131

Patriots' invasion of 30–31, 135–137
Quebec Act 114, 136
timeline 16
British imports. *See* nonimportation movement
British North America. *See also* American colonies
government and politics 91
history 35–80
British regulars. *See* Royal Army
British shipping 146, 156
British West Indies 91–94. *See also* West Indies
Brooklyn, New York 137
Broughton, Nicholson 150
Brown, Governor 152
Brown, John 148
Bunker Hill (Breed's Hill) 18, 116
Burgoyne, John 9, 19, 30, 139, 172

C

Cabot (Continental Navy) 152
Cahokia 65
Calvert, George 39
Calvinists 39, 124
Camden, South Carolina 142
Canada 29–30
Battle of Quebec 30, 137
British Canada 16, 30–31, 71, 103, 114, 131, 135–137
French Canada 15–16, 70, 179
Patriots' invasion of 135–137
Capture Act 153
Caribbean 28, 135, 179
Battle of the Saintes 19
West Indies 19, 33, 91–94, 145, 160, 162
Carleton, Guy 136, 154

Carteret, George 40
Carter, John 123
Catholics 39, 86, 114, 128
Cayugas 50, 51, 53
Charles II 41, 42
Charleston, South Carolina
British occupation 110
British seizure 32, 141, 163
rejection of Tea Act 113
timeline 18
Chaudierre River 30, 137
Cherokees 55, 56, 57–58, 68
Anglo-Cherokee war of 1759-1761 55
Cherokee War of 1776 56
Chesapeake Bay 143, 152
Chickamaugas 56, 57
Chickasaws 56, 63–65, 68
Chillicothe 59, 61
Chippewas 66, 68
Choctaws 65
Christians and Christianity 53, 67, 85, 86, 124
Cider Tax 128
"Circular Letter" (Boston legislature) 109
civil war 4
Clarke, John 42
Clark, George Rogers 61, 68
class distinctions 83–85
class warfare 132
Clinton, Henry 142, 163, 172
Coddington, William 42
Coercive Acts (Intolerable Acts) 17, 91, 111–117, 122, 177
Colombia 12
colonies
English. *See* American colonies

sugar colonies 91–92
Columbus (Continental Navy) 152
Committees of Correspondence 112, 118, 133
Committees of Inspection 130
Committees of Safety 53, 118
Common Sense (Paine) 1, 18, 118, 124, 127
commoners 133–134
communications 28, 121–123
Concord, Massachusetts 18, 116, 122–123
Congregationalism 53, 85, 124–125
Congress (Continental Navy) 152, 155
Congress of Augusta 57
Connecticut 31, 42–43, 141
    Stamp Act Congress 107
Connecticut Courant 123
conspiracy theory 113, 128–129
Constitution 20
Constitutional Convention 115, 178
Constitution of 1787 19, 132
constructivist perspective 8
contextual analysis 8
Continental Army 5, 119, 134–135, 144
    Battle of Brooklyn 137
    Battle of Bunker Hill 18, 116
    Battle of Long Island 18, 28, 138, 173
    Battle of Monmouth 19, 140
    Battle of Princeton 138
    Battle of Quebec 30
    Battle of Trenton 18, 138, 173
    Battle of Waxhaws 141
    Cherokee War of 1776 56
    defense of Boston 31
    regiments 120
    support for 118, 163
    tactical defeats 139
    timeline 18
Continental Congress 117, 133, 177
    Declaration of Independence 18
    First 17, 115, 126
    privateers 146
    Resolves 17
    Second 18, 116–117
    support for 118
    timeline 17, 18
Continental Navy 135, 146, 147
    Battle of Valcour Island 155
    converted merchants 152
    formation 151–153
    frigates 152
    losses 146
Conygham, Gustavous 156–157
Cornstalk 60–61
Cornwallis, Charles 172
    Battle of Monmouth 140
    southern campaign 142, 143, 163
    surrender 164
    timeline 19
Coshocton Indians 62–63
Covenant Chain 51
Cowkeeper (Ahaya of Cuscowilla) 58
Cowpens 142
Creeks 56, 57–58
cultural differences 4
currency 102
Currency Act 16, 91, 93–94, 102, 106, 111, 171
Cushing, Thomas 113
custom enforcement vessels 148

D

Deane, Silas 146, 151, 155–156, 160

de Beaujeu, Daniel 66
Declaration of Independence 117, 118
    excerpts 43, 126, 129
    press coverage 123
    publication 130
    signing 27
    timeline 18
Declaratory Act 16, 89, 91, 108, 171
Defense (Connecticut State cruiser) 150
Deganawidah 50
de Grasse, Comte (Joseph Paul) 143, 163
deism 124
DeLancey faction 41
Delaware (Continental Navy) 152
Delaware River 18, 138
Delawares 55, 61, 62, 63, 68, 103
Delaware state 31, 40, 107, 147
democracy 15
De Peyster, Arent 69
deprivation, relative 12
d'Estaing, Comte (Charles Hector) 160, 161–163
de Tocqueville, Alexis 25
Detroit, Michigan 61, 65
Dickinson, John 109
diplomacy, international 117
disease 45–46
Dominica 162
Dorchester Heights 137
d'Orvilliers, Comte (Louis Guillouet) 161
Dragging Canoe (Tsi'yugûnsi'ny) 55, 56
Drake (Continental Navy) 159
Dudingston, William 148
Dunmore, Lord 18
Dunmore's War 62

E

East Florida 58
East India Company 5, 17, 113, 114
economic development 11–12
economic inequality 12
economic resources 14
Edes, Benjamin 122
Edwards, Jonathan 125
Effingham (Continental Navy) 152
elites 41, 83, 133–134
emblems 11
England. *See also* Great Britain
    distance from colonies 28
English Channel 145, 156
English colonies
    American. *See* American colonies
    sugar colonies 91–92
English language 53
English merchants 5
enlightened paternalism 49
Enlightenment 124, 127
Enterprise 149
epidemiology 45–46
ethnic cleansing 48
ethnic conflict 13
ethnic identity 13
European immigrants 83
evangelicals 125
expansionism 46, 180

F

factionalism 129–130, 180
"father" role 65
Federalist Party 43
financial opportunities 14
First Continental Congress 115
First Great Awakening 124

Five Nations of the Iroquois.
    See Iroquois
flags 11
Florida 58
Fly (Continental Navy) 152
Forbes, John 120
foreign assistance 155–157
Fort Detroit 103
Fort Duquesne 16
Fort Frontenac 16
Fort Lee 18
Fort Loudon 55
Fort Niagara 52
Fort Ninety-Six 142
Fort Pitt 61, 103
Fort Randolph 60, 61
Fort Stanwix 59
Fort St. Johns 136
Fort St. Louis 59
Fort Ticonderoga 137, 149
Fort Washington 18
Fort William Henry 70
Founding Fathers 124
Fox, Charles James 143
Fox Indians 65
France 178, 179
    alliance with the Netherlands 139
    alliance with Spain 139
    coordination with Patriots 28, 29
    "father" role 65
    Indian alliances 64, 65–66, 69
    "the middle ground" 65
    naval intervention 160
    in North America 163
    Seven Years' War 46–47, 179–180
    support for the Patriots 124, 144, 145, 153, 155–156, 160
    Treaty of Alliance 19, 139
    Treaty of Paris of 1763 16
    in West Indies 33
Francophiles 51
Franklin, Benjamin
    and Conygham 156–157
    in France 146, 155–156, 160
    leadership 40, 99, 114, 133
    on the question of independence 130
    Poor Richard's Almanac 159
Franklin, William 40
Freeman's Farm 139
French-American army 143
French and Indian War 37, 69–71, 120
    Battle of Fort William Henry 70
    Battle of Quiberon Bay 70
    timeline 15–16
French Canada 15–16, 70, 179
French Navy 28, 33, 92, 143, 160
    Battle of Quiberon Bay 70
    Battle of the Saintes 19
    Battle of Ushant 160–161
    Battle of Yorktown 163–165
    Brest Fleet 161
    capture of Dominica 162
    capture of Grenada 162
    capture of St. Vincent 162
    coordination with Patriots 135
    d'Estaing's deployment 161–163
    in North America 163
    tactics 160
frigates 152
frontier conditions 28
frontier fighters 120
fur trade 65

G

Gadsden, Christopher 151
Gage, Thomas 54, 115, 172
Galloway, Joseph 40
Gaspee affair 17, 112, 148
Gates, Horatio 54, 142, 154, 160
GDP (gross domestic product) 11–12
geography 27–33, 120
George III 71, 107
    American sentiments toward 125–126
    Gaspee affair 148
    Indian support for 43, 52, 64
    Stamp Act 105
    timeline 16
Georgia 32, 38
    Loyalists 118
    Native Americans in 57, 114
    Patriot liberation of 143
German Protestants 83
Germantown, Pennsylvania 139, 173
Gerrish, Edward 111
gift giving 70, 103
Gill, John 122
Glorious Revolution of 1688-1689 41
Goddard, Sarah 123
Goddard, William 123
Good Peter (Peter Agworondougwas) 53
government and politics 13–15, 89–94
    countermeasures to insurgency 169–173
    transition from insurgency to governance 177–179
Graves, Thomas 143, 164
Graymont, Barbara 44, 47
Great Awakening 124, 125
Great Britain 4, 101. *See also* England

    Anglo-Cherokee war of 1759-1761 55
    British North America 35–80, 91
    British West Indies 91–94
    Capture Act 153
    Cider Tax 128
    countermeasures 169–173
    Currency Act 16, 91, 93–94, 102, 106, 111
    Declaratory Act 16, 91, 108
    French and Indian War 15–16, 37, 69–71, 120
    government and politics 89–94
    House of Commons 5
    House of Lords 5
    Impartial Administration of Justice Act 114
    Indian allies 28, 51, 63
    initial response to insurgency 171–172
    Intolerable Acts (Coercive Acts) 17, 91, 111–117, 122, 177
    invasion of St. Lucia 162
    legislative acts 171
    Massachusetts Regulatory Act 114
    military response 172–173
    Molasses Act 92
    national debt 49, 71, 92, 171
    Navigation Acts 148
    northern campaigns 137–141
    occupation of Boston 31, 172
    occupation of the colonies 114, 123
    piracy act 146
    political response 172–173
    Pontiac's War 49, 66, 67, 103–104
    Proclamation of 1763 6, 16, 39, 48, 50, 104, 129, 131
    Quartering Act 16, 108, 114
    Quebec Act 114, 136

Seven Years' War 46–47, 49–50, 179–180
southern campaigns 141–144, 163
Stamp Act 16, 41, 91, 102, 104, 105
success as a world empire 179–180
Sugar Act 16, 91, 92–93, 102, 106
sugar colonies 91–92
Suspending Act 109
taxation of the colonies 15, 92–94
Tea Act 17, 111–117
timeline 16
Townshend Acts 17, 108–110
trade and navigation laws 101
Treaty of Paris of 1763 16
Treaty of Paris of 1783 19, 56, 69, 144, 164
view of Indians 49–50
Great Lakes 30, 70
Great Lakes region. *See pays d'en haut (upper country)*
Great League of Peace and Power 50
greed 13
Greene, Nathanael 5, 140, 142, 143, 144
Green Mountain Boys 149, 153–154
Grenada 162
Grenville, George 102, 104, 105, 107
gross domestic product (GDP) 11–12
guerre de course (war of the chase) 145–147
guerre de escadre (war of the squadron) 145, 147
Guilford Courthouse (North Carolina) 143
Guillouet, Louis (Comte d'Orvilliers) 161

H

Hacker, Hoysted 158
Halifax, Nova Scotia 18
Hamilton, Henry 68
Hancock (Continental Navy) 152
Hancock, John 114
Handsome Fellow 57
Hannah (American merchant) 148
Hannah (Washington's Navy) 150
Harvard University 124
heathenism 124
Hector, Charles (Comte d'Estaing) 160, 161–163
Henry, Patrick 81, 106
Hessian mercenaries 18, 54, 119, 138
Hewes, Joseph 151
Hiawatha 50
historical context 9–11, 35–80
  Patriot insurgency course 101–117
  timeline 15–20
HMS *Bolton* 152
HMS *Carleton* 155
HMS *Cerebus* 158
HMS *Gaspee* 17, 112, 148
HMS *George* 149
HMS *Hawke* 152
HMS *Inflexible* 155
HMS *Prince of Orange* 156
HMS *Serapis* 159
Holt, John 123
Hopkins, Esek 152
Hopkins, Stephen 148, 151
Hornet (Continental Navy) 152
hostages 152
Howe, William 31, 137, 139, 172
Hudson River 18
Hudson Valley 30
Hudson Valley-Lake Champlain corridor 9, 29–30, 149
  map 30

timeline 19
Huron-Petuns 65
Hurons 63, 66, 68, 103
Hutchinson, Anne 42
Hutchinson, Thomas 106, 112–113, 130

I

identity, ethnic 13
ideology 123–131
Illinois Indians 59, 65
Impartial Administration of Justice Act 114
imperialism 105
indentured servants 18
independence 130–131
Indian nations 20. *See also* Native Americans
Indian Territory 47
informers and spies 53
infrastructure 28–29
Ingersoll, Jared 104
insurance rates 146, 156
insurgency. *See also* revolution
    causes of 179–180
insurgent strategy 138
international diplomacy 117
Intolerable Acts (Coercive Acts) 17, 91, 111–117, 122, 177
Irish immigrants 83
Irish Sea 156
Iroquois 44, 50
    alliance with Royalists 131
    Battle of Oriskany 53
    Covenant Chain 51
    military organization 51
    raids against Patriots 140
    Treaty of Fort Stanwix 55, 59
Iroquois Confederacy 50, 53
irregular forces 5, 120
irregular warfare 144
Irving, William 140

J

Jamaica 33, 91–92
James II 40, 41, 128
James River 163
Jamestown, Virginia 38
Jefferson, Thomas 5, 56, 126, 133
Johnson, William 51
Jones, John Paul 157–159

K

Kaskaskia 65
Kentucky 59, 60, 61
Keppel, Admiral 161
Kickapoo 65, 103
Killbuck 62–63
King's Mountain, North Carolina 142
kinship systems 66
Kirkland, Samuel 52
Kispokis 59
Knox, Henry 137

L

Lafayette, Marquis de 140, 163, 164
Lake Champlain
    Battle of Valcour Island 155
    first campaign 149–150
    second campaign 153–155
Lake Champlain-Hudson Valley corridor 9, 29–30, 149
    map 30
    timeline 19
Lake Champlain Valley 70
land 129

land operations 5, 134–135
   analysis 144
     northern campaigns 137–141, 144
     southern campaigns 141–144, 163
land taxes 40
Langdon, John 151
Lee (Massachusetts State) 150
Lee, Arthur 57
Lee, Charles 56, 140
Lee, Harry (Light Horse) 140
Lee, Richard Henry 151
Lesser Antilles 33
Letters from a Farmer in Pennsylvania (Dickinson) 109
Lewis, Andrew 59
Lexington (Continental Navy) 151
Lexington, Massachusetts 18, 116, 122–123
liberation theology 67
liberty 126–127, 132
Liberty 149
Light Horse (Harry Lee) 140
Little Abraham (Tigoransera) 51
Little Carpenter (Attakullakulla) 55
Livingston faction 41
local or regional movements 11
Locke, John 127
Long Island 18, 28, 138, 173
looting 13, 159
Lord Dunmore's war 59, 60
Lower Cherokees 56
Lower Creeks 57
Loyalists 4, 29, 32, 118, 130
   Battle of Moore's Creek 38
   Battle of Oriskany 53
   Battle of Waxhaws 141
   colonial population 132
   counterinsurgents 132
   hatred for Patriots 141, 142
   motivation and behavior 132
   newspapers 121
   refugees 133
   southern campaigns 141–144
lumber trade 42

M

Maclay, Edgar 147
Mahican 53
Maine 31, 41, 153
Manhattan 18
Maquachakes 59, 60–61
Marine Committee 151–153
Marion, Francis 9
maritime tradition 147
maritime war 148
   combatants 147–148
   Lake Champlain campaigns 149–150, 153–155
maroons 91
Martha's Vineyard 41
Martinique 150, 162
Maryland 31, 39, 107
Maryland Toleration Act 39
Mascouten 65, 103
Mason, John 42
Massachusetts 31, 41
   praying towns 53
   privateers 146
   Stamp Act Congress 107
   timeline 17
Massachusetts Bay 41
Massachusetts Regulatory Act 114
Massachusetts Spy 122–123
McDougall, Alexander 151
McGillivray, Alexander 58

Mediterranean 145, 160
Mein, John 122
Menominee Indians 65
mercantilism 102
merchant marine 101
Mesoamerica 45
Methodists 125
Miamis 65, 67, 68, 103
Michilimackinac 65
Middle Colonies 27, 31
    campaigns 137–141
    ports 29
    religion 86
"the middle ground" 65
middling classes 132
middling sort 83
military occupation 110–111, 114
military planning 28
military power 14
militias 116, 119, 142
Mingos 61, 63, 66, 68, 103
Minutemen 18, 116
mission civilisatrice 49
missionaries 53
Mississippi 63
Mississippi River 65, 179
mob violence 106, 107
modernization 12
Mohawk River 30
Mohawks 50, 51–52, 53, 55
Mohawk Valley 30
Mohicans 53
Molasses Act 92
Monmouth, New Jersey 19, 140
Montgomery (Continental Navy) 151, 152
Montgomery, Hugh 111

Montgomery, Richard 135–137, 154
Montreal, Canada 30, 71, 136
Moore's Creek 38
Morgan, Daniel 137, 142, 144
Morgan, George 61
Morris, Robert 151
mourning wars 50

N

names 11
Nancy 150
Nanticokes 55
Nantucket 41
Narragansett Bay 148
Nassau, Bahamas 152
nation(s) 13
Native Americans 43–69, 115, 180
    Battle of Valcour Island 155
    British view of 49–50
    French and Indian War 15–16, 37, 69–71, 120
    Great League of Peace and Power 50
    mourning wars 50
    objective 4
    Pontiac's War 49, 66, 67, 103–104
    pro-British Indians 28, 51, 131
    pro-Patriot Indians 52
    Stockbridge Indians 54
    timeline 20
    view of American Revolution 43, 51–52
nativists 67–68
natural resources 13
Naval Committee/Marine Committee 151–153
naval operations 5, 144–164
    battle of Chesapeake Bay 143

forms of warfare 145
naval superiority 163
Navigation Acts 148
Neolin 67
the Netherlands 139, 153
neutrality 130
New England 31
   campaigns 137–141
   colonies 27
   merchant marine 101
   Patriot insurgency 31, 118
   ports 29
   religion 53, 85–86
New France 65
New Hampshire 20, 31, 42
New Jersey 31, 40, 107, 118, 147
New Lights 125
Newport, Rhode Island 42, 161
newspapers 121–123
New York City, New York 29, 31, 163
   Battle of Brooklyn 137
   British occupation 110, 123
   rejection of Tea Act 113
New York Journal 123
New York state 31, 41
   Native Americans 51
   rejection of Quartering Act 108
   Stamp Act Congress 107
   Suspending Act 109
Nimham, Daniel 54
nonimportation movement 110, 114, 130
   Boston Non-Importation Agreement 17
   non-importation resolutions 171
   timeline 17
North Carolina 32, 38
   Loyalists 118

Patriot militias 142
northern Atlantic Ocean 145
northern campaigns 137–141, 144
Nova Scotia 41

O

Oconostota 56
Odawa 103
Oglethorpe, James 38
Ohio Valley
   Dunmore's War 62
   Indians 20, 59, 62, 103, 131
Ojibwe 65, 103
Old Lights 125
Old Smoke (Sayengeraghta) 52
"Olive Branch Petition" 116
Oneidas 50, 51, 52–53, 53
Onondagas 50, 51, 53
Onontio 66, 67, 68
Oriskany, New York 53
Otis family 106, 130
Ottawas 65, 68
   Cherokee War of 1776 55
   Pontiac's War 66, 67, 103
Overhill Cherokees 56

P

paganism 124
Paine, Thomas 35
   Common Sense 1, 18, 118, 124, 127
   timeline 18
Palliser, Hugh 161
paramilitary operations 134–144
Parker, John 116
partisanship 129–130
paternalism 49, 65, 105
Patriot insurgency 99–168

armed component 118–121
auxiliary forces 5, 117–118
Boston Tea Party 113
causes 179–180
command and control 117–123
Committees of Inspection 130
Committees of Safety 118
commoners 133–134
communications network 112
coordination with French 28, 29
course 101–117
Declaration of Independence 27, 118
elites 133–134
expansion into global war 162
experience and education 121
external support 153
factions 4
features similar to other resistance movements 4
First Continental Congress 115
foreign assistance 124, 153, 155–157, 160
formation 71
four major problems 4
government countermeasures 169–173
hatred for Loyalists 141, 142
historical context 35–80
ideology 123–131
Indian support 52
initial response 171–172
invasion of Canada 30–31, 135–137
irregular forces 5, 120
land operations 5, 134–135, 144
leadership 117–123, 133
lessons from 5
maritime campaign 144–148

maritime combatants 147–148
means of resistance 130
militias 116, 119, 142
motivation and behavior 131–134
nature of resistance 119–120
naval campaigns 5
Naval Committee/Marine Committee 151–153
newspapers 121–123
nonimportation agreements 17, 110, 114
northern campaigns 137–141, 144
objective 5
"Olive Branch Petition" 116
operations 134–164
organizational structure 117–123
paramilitary operations 134–144
"Petition to the King" 115
public component 121–123
at sea 144–148, 156
Second Continental Congress 18, 116–117
sentiments toward George III 125–126
shadow government 118
southern campaigns 141–144
spark that gave rise to 102
timeline 15–20
transition to governance 177–179
underground component 117–118
unique features 4
Paul, Joseph (Comte de Grasse) 143, 163
Paulus Hook 141
Paya Mataha 64
pays d'en haut (upper country) 49, 65–68, 70, 103, 131
Pearson, Richard 159

Penn Family 40
Pennsylvania 31, 39–40, 59, 83
    Stamp Act Congress 107
Pennsylvania Evening Post 123
Pennsylvania Journal 122
Penn, William 39
Pensacola, Florida 65
Percy, Earl 116
Philadelphia, Pennsylvania 31, 39
    British capture 139
    British occupation 122, 123
    British withdrawal 140
    First Continental Congress 115
    rejection of Tea Act 113
    timeline 19
physical environment 8–9, 25–34
Piankashaw 103
Piquas 59, 60, 61
piracy 146
Pitcairn, John 116
Pittsburgh, Pennsylvania 16
Pitt, William the Elder 15, 70–71, 108, 127
Plymouth Colony 41
politics 4, 12, 13–15, 89–94
Pontiac 103
Pontiac's War 49, 66, 67, 103–104
Poor Richard's Almanac (Franklin) 159
population 83–86, 171
ports 28, 29
Portsmouth, Rhode Island 42
positivist perspective 8
postmillennialism 124
Potawatomis 65, 66, 67, 68, 103
poverty 11–12
praying towns 53
Presbyterians 52, 86, 125

Preston, Thomas 111
Princeton, New Jersey 19, 138, 173
prisoners of war 152
privateers 5, 135, 146, 147, 148
    state 150–151
    sustaining 155–157
private property 127
propaganda 141, 172
property rights 127
Protestants and Protestantism 39, 123, 124–125, 128
    German Protestants 83
Providence-12 (Continental Navy) 152, 158
Providence-28 (Continental Navy) 152
Providence Gazette 123
Providence, Rhode Island 42
public communications 121–123
Puritans 39, 41, 42, 53, 124

Q

Quakers 39, 85, 86
Quartering Act 16, 108, 114
Quebec Act 114, 136
Quebec City, Canada 30, 65
    Battle of Quebec 30, 137
    British capture of 71
Quebec Province 114
Quiberon Bay 70

R

Raleigh (Continental Navy) 152
Randolph (Continental Navy) 152
Ranger (Continental Navy) 158
rebellion
    poverty-related indicators 12
    risk factors for 11

refugees
   Loyalist 133
   Native American 65
regime types 14–15
regional movements 11
relative deprivation 12
religion 85–86
   in the colonies 27, 39, 53
   influences of 123–125
religious freedom 42
Reprisal (Continental Navy) 155
Resolves
   timeline 17
   Virginia Resolves 106
Revenge (Continental Navy) 156–157
Revere, Paul 115
revivalism 125
revolution
   causes or bases 3
   definition 3
   political factors 14
Rhode Island 31, 42, 107, 110, 148
right to own property 127
risk factors for conflict 12
roads 28
Rochambeau, General 143, 163
Rockingham, Marquess of 108
Roman Catholics 39, 86, 114, 128
Roosevelt, Theodore 157
rough terrain 8–9, 28, 120
Royal Army 119
   Battle of Bunker Hill 18, 116
   Battle of Fort William Henry 70
   Battle of Long Island 18, 28, 138, 173
   Battle of Monmouth 19
   Battle of Oriskany 53

Battle of Quebec 30
Battle of Trenton 18, 138, 173
Battle of Waxhaws 141
Battle of Yorktown 143–144
Battles of Lexington and Concord 18, 116
   occupation of the colonies 110–111
Royalists 130, 131
Royal Navy 28, 29, 33, 92, 130, 143, 145, 157
   Battle of Brooklyn 137
   Battle of Quiberon Bay 70
   Battle of the Saintes 19
   Battle of Ushant 160–161
   Battle of Valcour Island 155
   Battle of Yorktown 163–165
   Capture Act 153
   Channel Fleet 161
   combined state navies and privateer forces against 150
   custom enforcement vessels 148
   French and Indian War 16
   Gaspee affair 17, 112, 148
   Lake Champlain force 155
   losses 146
   surrender of *Serapis* 159
   tactics 160
   timeline 16
   war of the squadron (*guerre de escadre*) 147
Royal Proclamation of 1763 6, 39, 48, 104, 131
   colonist views 50, 129
   timeline 16
rural commoners 133
Russia 153, 178

S

Saint-Domingue 91
Saratoga, New York 19, 28, 139
Sauk 65
Savannah, Georgia 19, 32, 56, 141, 162
Sayengeraghta (Old Smoke) 52
Schuyler, Philip 154
Scotch-Irish population 83
Second Continental Congress 18, 116–117
Second Great Awakening 124
sectional cleavages 4
Selkirk, Earl 158
Seminoles 56, 58–59
Senecas 50, 51, 63, 68
    Battle of Oriskany 53
    Pontiac's War 103
Seven Years War. *See also* French and Indian War
Seven Years' War 46–47, 49–50, 69, 146, 179–180
shadow government 118
Shawnees 59–63, 68, 69, 103
    Cherokee War of 1776 55
    Pontiac's War 66
    Treaty of Fort Stanwix 59
Sherman, Philip 42
ship building 102
shipping 101, 147
shot heard 'round the world 18
Sinhalese 12, 13
Six Nations 59
slaves and slavery 37, 38, 177–179
    African slaves 91
    black slaves 83, 85
    as property 127
    Three-Fifths Compromise 179

timeline 18
smallpox 55
Smith, Francis 115
smuggling 147, 171
social classes 83–85
social contract theory 127
social movements
    definition 14
    historical context 10
socioeconomic conditions 11–13, 81–87
socioeconomic theories 11
Sons of Liberty 109
South Carolina 32, 37
    British campaigns 142
    Cherokee War of 1776 56
    Council of Safety 57
    Loyalists 118
    Patriot liberation 143
    physical environment 9
    Stamp Act Congress 107
    Yamassee War of 1715 59
southern campaigns 141–144, 163
Southern Colonies 32
    map 27, 32
    religion 86
Spain 28, 178
    alliance with France 139
    pro-Spanish indians 58
    support for the Patriots 145, 153
spiritualism, native 67
Sri Lanka 12, 13
Stamp Act 91, 102, 104, 105, 171
    Patriot resistance 41
    timeline 16
Stamp Act Congress 16, 107
Stamp Act Crisis 104–108

Star-Spangled Banner 158
Starved Rock 59
state war debts 178
Stirling, Lord 152
St. Lawrence Valley 65, 137
St. Lucia 162
St. Mary's Isle 158
Stockbridge Indians 54
Stockbridge, Massachusetts 53–54
Stony Point 141
Stuart, Henry 55
Stuart, John 48, 63, 64, 128
St. Vincent 162
Sugar Act 16, 91, 92–93, 102, 106, 171
sugar colonies 91–92
sugar islands 28, 91, 92
Sullivan, John 140
Surprise (Continental Navy) 156
Suspending Act 109
symbols 11

T

Taitt, David 57
Taliban 9
Tamils 12, 13
Tarleton, Banastre 9, 141, 142
taxes and taxation
    Cider Tax 128
    "Circular Letter" on 109
    on colonies 92, 105
    as constitutional issue 93
    custom enforcement vessels 148
    external taxes 108
    internal taxes 108
    land taxes 40
    resistance to 41, 71
    Stamp Act 102, 104, 105

Tea Act 111–117
timeline 15
Townshend duties 17, 108–110
Tea Act 17, 111–117
Tennessee 63
terrain. *See* physical environment; *See* rough terrain
Thawekila 59
Thayendanegea (Joseph Brant) 51–52, 140
Thomas, Isaiah 122–123
Three-Fifths Compromise 179
Tigoransera (Little Abraham) 51
Tilly, Charles 10
timeline 15–20
Tonyn, Patrick 58
Tories 123. *See also* Loyalists
Towne, Benjamin 123
Townshend Acts 17, 108–110, 171, 172
Townshend, Charles 108, 109
trade routes 103
treaties 117
Treaty of Albany 54
Treaty of Alliance 19
Treaty of Dewitt's Corner 56
Treaty of Fort Stanwix 55, 59
Treaty of Paris of 1763 16
Treaty of Paris of 1783 19, 56, 69, 144, 164
Trenton, New Jersey 18, 138, 173
Trumble (Continental Navy) 152
Tryon, William 141
Tsi'yugûnsi'ny (Dragging Canoe) 55, 56
Tuscarora 50, 51, 52, 53
Two Treatises of Government (Locke) 127
Tybee Island 162

U

Uhhaunauwaunmut, Solomon 54
United States Constitution 19, 20
United States Constitution of 1787 132
United States Navy. *See also* Continental Navy
   father 157–159
   formal recognition 158
United States of America
   alliance with France 139
   Articles of Confederation 19, 177
   Declaration of Independence 18, 43
   Indian allies 63
   thirteen colonies 27, 37–43
   Treaty of Alliance 19
   Treaty of Paris of 1783 19, 56, 69, 144, 164
   war debt 178
Unity 150
upper country (*pays d'en haut*) 49, 65–68, 70, 103
Upper Creeks 57
urban elites 133
Ushant 160–161

V

Valcour Island 155
Valley Forge, Pennsylvania 19, 140
Van Cortlandt Park 54
vandalism 114
Viet Cong 9
Vietnam 9
Vincennes, Michigan 65
violence
   domestic political violence 11
   mob violence 106, 107
   profligate 133
   socioeconomic theories for 11
   transition to 10
Virginia (Continental Navy) 152
Virginia 32, 38–39
   Committee of Correspondence 112
   House of Burgesses 112
   Loyalists 118
   Native Americans 47
   Patriot insurgency 118
   timeline 18
Virginians 56
Virginia Regiment 119, 120
Virginia Resolves 106
Virginia Stamp Act Resolutions 16
von Steuben, Friedrich 140

W

Walpole, Robert 102
Wappingers 53, 54
war debt 178
war of the chase (*guerre de course*) 145–147
war of the squadron (*guerre de escadre*) 145, 147
Warren (Continental Navy) 152
Warwick, Rhode Island 42
Washington (Continental Navy) 152
Washington, George 5, 133
   Battle of Brooklyn 137
   as Commander in Chief 18, 19, 31, 116, 120, 134–135, 137
   experience and education 121
   as frontier fighter 120
   insurgent strategy 138
   leadership 133, 139, 140
   retreat across the Delaware River 138
   as strategist 144, 163–164

timeline 18, 19
Washington's Navy 148, 150, 150–151
Wasp (Continental Navy) 152
waterways, navigable 28
Waxhaws massacre 141
Wayne, Anthony 140
Wea 103
Wentworth, Benning 42
West Florida 57
West Indies 33, 145, 160, 162
    Battle of the Saintes 19
    British West Indies 91–94
Whately, Thomas 104, 112–113
Whigs 70, 127, 128
Whipple, Abraham 148
White Eyes 62–63
Whitefield, George 125
White Haven, England 158
White, Hugh 111
White, Richard 65, 68
Wickes, Lambert 155, 157
Williams, Roger 42
Winnebago Indians 65
Wolfe, James 16
Wooster, David 136
Wyandots 61, 65, 67, 68, 103
Wyatt, Francis 47

Y

Yale University 124
Yamassee War of 1715 59
Yorktown, Virginia 143–144, 163–165, 173

www.ingramcontent.com/pod-product-compliance
Lightning Source LLC
Chambersburg PA
CBHW051752100526
44591CB00017B/2664